Critters

Authors
Maureen Murphy Allen
Betty Cordel
Debby Deal
Suzy Gazlay
Denise Del Grosso Gilliland
Carol Gossett
Gale Philips Kahn
Myrna Mitchell
Michelle Pauls
Suzanne Scheidt
Vincent Sipkovich

Editors
Michelle Pauls
Betty Cordel

Illustrator
Dawn DonDiego

Desktop Publisher
Roxanne Williams

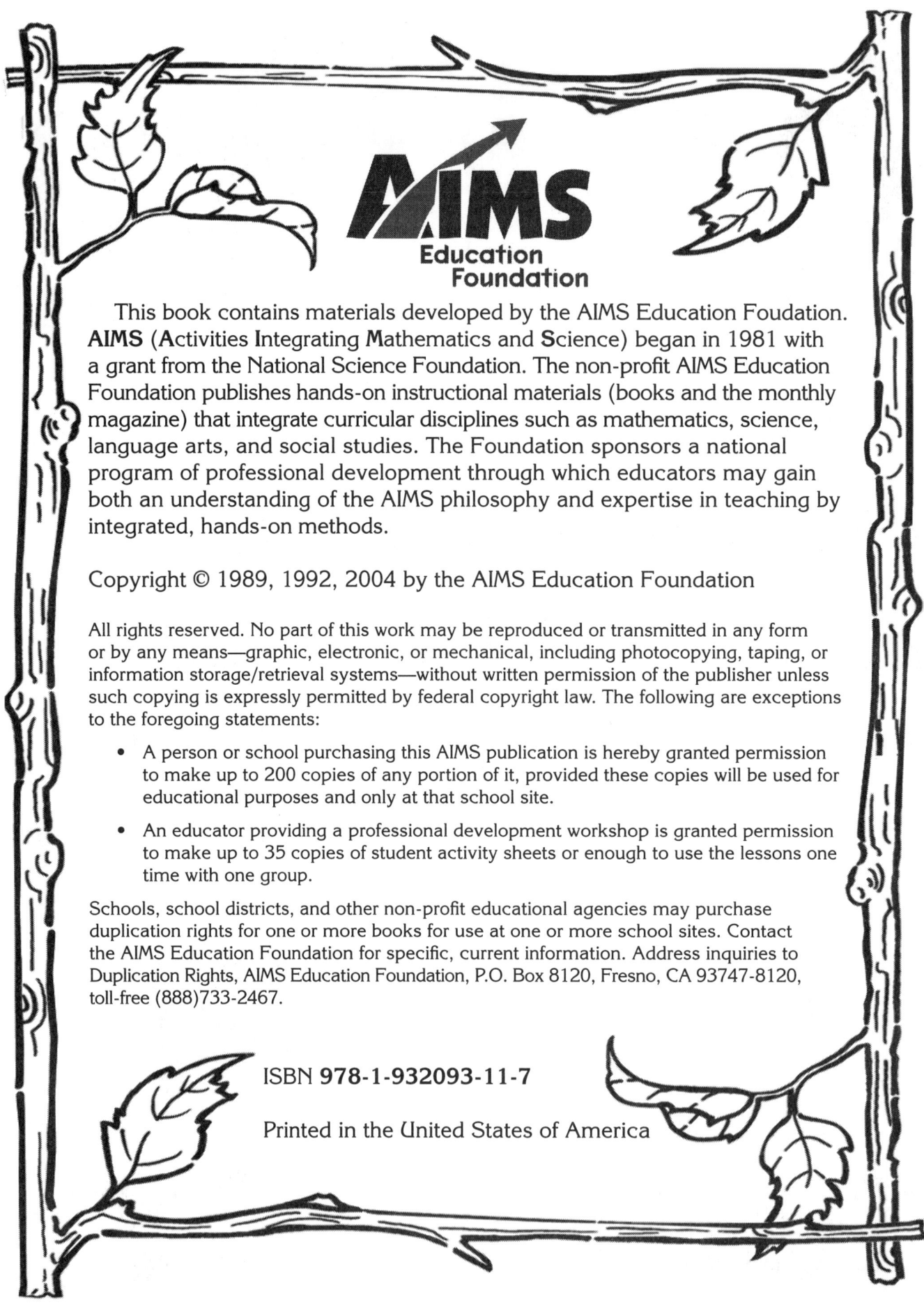

This book contains materials developed by the AIMS Education Foudation. **AIMS** (**A**ctivities **I**ntegrating **M**athematics and **S**cience) began in 1981 with a grant from the National Science Foundation. The non-profit AIMS Education Foundation publishes hands-on instructional materials (books and the monthly magazine) that integrate curricular disciplines such as mathematics, science, language arts, and social studies. The Foundation sponsors a national program of professional development through which educators may gain both an understanding of the AIMS philosophy and expertise in teaching by integrated, hands-on methods.

Copyright © 1989, 1992, 2004 by the AIMS Education Foundation

All rights reserved. No part of this work may be reproduced or transmitted in any form or by any means—graphic, electronic, or mechanical, including photocopying, taping, or information storage/retrieval systems—without written permission of the publisher unless such copying is expressly permitted by federal copyright law. The following are exceptions to the foregoing statements:

- A person or school purchasing this AIMS publication is hereby granted permission to make up to 200 copies of any portion of it, provided these copies will be used for educational purposes and only at that school site.

- An educator providing a professional development workshop is granted permission to make up to 35 copies of student activity sheets or enough to use the lessons one time with one group.

Schools, school districts, and other non-profit educational agencies may purchase duplication rights for one or more books for use at one or more school sites. Contact the AIMS Education Foundation for specific, current information. Address inquiries to Duplication Rights, AIMS Education Foundation, P.O. Box 8120, Fresno, CA 93747-8120, toll-free (888)733-2467.

ISBN 978-1-932093-11-7

Printed in the United States of America

I Hear and I Forget,
I See and I Remember,
I Do and I Understand.

-Chinese Proverb

Table of Contents

Introduction v
Critters Vocabulary vi
Critters Literature.................................... vii

The Critter Connection
Rubber Band Books: Reading
  in the Content Area 1
Silkworms 2
Mealworms 4
Insects...6
Spiders 8
Frogs and Toads......................................10
Camouflage 12
Food Chains 14

Critter Observation
Lenses & Ladybugs * 16
Spider Spoofs and Proofs * 22
Spying on Spiders * 29
Growing Pains * 34
Mealworm Moments, Part One * 40

Critter Behaviors
Mealworm Moments, Part Two * 48
Move Along, Mealworm * 56
Mealworm Hop (Song) 59
Web Threads 61
Goldfish Gulps * 65
Hot Foot, Cold Feet * 68

Critter Classification and Characteristics
Wings 'n' Webs 74
Animal Antics 81
Fishing for Fins..90
Beetle Mania...96
A Look at Lepidoptera........................... 106
Frog and Toad are Kin........................... 118

Critter Camouflage and Adaptations
Under Cover 125
Wonderful Webbed Feet........................ 131
Critters Hide 'n' Seek 136
Gone Fishing 140
Missing Moths....................................... 145
Table Manners 152
I'm Stuck on You 158

Critter Life Cycles
Mealworms on Stage * 166
This is Your Life, Tadpole * 171

Critter Interdependence
Food Chain.. 181
Chain Games.. 187
Catch Me if You Can 193
Census Takers 197
Biome Boxes 203
Who's Home in the Biome? 214

* Denotes activities that use live critters.

Introduction to the Revised Edition

Since its original publication in 1989, *Critters* has been one of AIMS' most popular books. The appeal of critters in the classroom is nearly universal, especially when they are the vehicle for gaining a deeper understanding of important math and science concepts. Now, after 15 years in print, *Critters* has undergone a major revision.

Besides adding more than 10 new activities, each of the remaining original activities has been updated to match our current format, including alignment with the latest national standards in math and science. There are also Internet and literature connections, and a collection of rubber band books that provides opportunities for reading in the content area.

Another way in which the revised *Critters* differs from the original is in its organization. While the original was organized by critter, this edition focuses on six main themes: *Observation, Behaviors, Classification and Characteristics, Camouflage and Adaptations, Life Cycles,* and *Interdependence.* Each of these themes is supported by anywhere from two to seven activities that use a variety of critters.

Within each section, the activities are generally arranged starting with those appropriate for lower grades, and moving to activities more suited for upper grades. In virtually every case, the activities can be modified up or down to suit the needs of your students.

Critters Vocabulary

Abdomen: the last section of an insect's three main body parts, or the last of a spider's two body parts

Adaptation: a change made to fit the environment

Antennae: (singular, antenna) the sensory organs of an insect used to smell, taste, feel, and sometimes hear

Biome: a large area (such as the ocean) that has similar geography, climate, plants, and animals

Camouflage: an organism's ability to blend in with the environment

Carnivore: an animal that eats only meat (other animals)

Census: the counting of a population

Cephalothorax: the first of two body sections in spiders and some crustaceans consisting of the head and thorax fused together

Chrysalis: another name for the pupal stage of moths, butterflies, and other insects that have a similar metamorphosis

Cold-blooded: in animals, the lack of ability to maintain a constant body temperature independent of the outside temperature. (Insects, reptiles, fish and amphibians are cold-blooded.)

Cocoon: the pupal case of moths and butterflies, made out of a silk-like material

Consumer: any animal or plant that gets its energy by eating other animals or plants

Gills: organs in fish that separate dissolved oxygen from water

Habitat: the environment in which an animal lives

Herbivore: an animal that eats only plants

Invertebrate: an animal without a backbone

Larva: (plural, larvae) the second stage of an insect that goes through complete metamorphosis (egg, larva, pupa, adult). Insect larvae look very different from the adults. For example, moth or butterfly larvae are caterpillars, fly larvae are maggots, and beetle larvae are grubs.

Life cycle: the stages of development through which an organism passes

Metamorphosis: the process by which many insects change from egg to adult. In *complete metamorphosis*, the insect goes through four stages: egg, larva, pupa, and adult. In *incomplete metamorphosis*, the insect goes through three stages: egg, nymph, and adult.

Molt: to shed the outer skin or exoskeleton. As an insect grows, it sheds its skin several times before it reaches the adult stage.

Nymph: the immature stage of an insect that goes through incomplete or simple metamorphosis. The nymph looks like the adult but is much smaller.

Omnivore: an animal that eats both plants and animals

Population: the total number of organisms of one species in a particular area

Predator: an animal that hunts, kills, and eats other animals

Prey: an animal that is caught, killed, and eaten by another animal (predator)

Producer: green plants that are able to make their own food

Pupa: (plural, pupae) the inactive stage of an insect that goes through complete metamorphosis. The larva changes into a pupa before becoming an adult.

Spider: an arthropod with two main body parts, 8 legs, a silk-producing organ, and fangs

Thorax: the middle section of an insect's body to which the legs and wings are attached

Vertebrate: an animal with a backbone

Warm-blooded: in animals, the ability to maintain a constant body temperature independent of the outside temperature. (Birds and mammals are warm-blooded.)

Critters Literature

Primary Grades Fiction

Insects
Brown, Ruth. *Ladybug, Ladybug*. Puffin. New York. 1992.
(Beautiful illustrations accompany the familiar ladybug nursery rhyme.)

Carle, Eric. *The Grouchy Ladybug*. Scott Foresman. Chicago. 1996.
(The grouchy ladybug learns to be nicer, happier, and better behaved.)

Hong, Lily Toy. *The Empress and the Silkworm*. Albert Whitman & Co. Morton Grove, IL. 1995.
(A fictionalized account of the Empress of China's discovery, around 2700 BC, that the cocoons of the worms in her mulberry trees were made of a fine, shiny silken thread that could be made into beautiful cloth.)

Reptiles and Amphibians
Carle, Eric. *The Mixed-Up Chameleon*. Harper Trophy. New York. 1988.
(The chameleon is bored with life until he discovers that he can change his shape and size as well as his color.)

Spiders
Carle, Eric. *The Very Busy Spider*. Putnam Publishing. New York. 1995.
(Through the illustrations, the sequence of how a web is built is portrayed in this book. A board book edition of the story features a raised spider web children can actually feel.)

Primary Grades Non-fiction

Adaptations and Camouflage
Arnosky, Jim. *I See Animals Hiding*. Scholastic, Inc. New York. 2000.
(Using beautiful watercolors, the author shows the camouflage that various animals use.)

Goodman, Susan. *Claws, Coats, and Camouflage*. Millbrook Press. Brookfield, CT. 2001.
(Discusses adaptation using four broad categories: "Fitting In," "Staying Safe," "Getting Food," and "Making a New Generation.")

Food Chains and Webs
Lauber, Patricia. *Who Eats What?* HarperCollins. New York. 1995.
(Explains the concept of a food chain and how plants, animals, and humans are ecologically linked.)

Insects
Allen, Judy. *Are You a Butterfly?* Kingfisher Publishing. New York. 2000.
(With simple text and colorful drawings, introduces the life cycle of a butterfly, showing how it changes from an egg to a caterpillar, to a chrysalis, to an adult butterfly.)

Allen, Judy. *Are You a Ladybug?* Kingfisher Publishing. New York. 2000.
(With simple text and colorful drawings, introduces the life cycle of a ladybug, showing how it changes from an egg to an adult ladybug.)

Cassie, Brian and Jerry Pallotta. *The Butterfly Alphabet Book*. Charlesbridge Publishing. Boston. 1995.
(Informative, simple text accompanies beautiful paintings of butterflies that have names beginning with every letter of the alphabet. Also discusses the life cycle of a butterfly.)

Pallotta, Jerry. *The Icky Bug Alphabet Book*. Charlesbridge Publishing. Boston. 1993.
(Informative, simple text accompanies pictures of ladybugs, spiders, butterflies, and other "bugs.")

Reptiles and Amphibians

Gibbons, Gail. *Frogs*. Holiday House. New York. 1994.
(The life cycle of frogs is explained using simple language and is accompanied by illustrations.)

Pallotta, Jerry. *The Frog Alphabet Book*. Charlesbridge Publishing. Boston. 1990.
(Informative, simple text accompanies realistic pictures of frogs, salamanders, and other amphibians.)

Pallotta, Jerry. *The Yucky Reptile Alphabet Book*. Charlesbridge Publishing. Boston. 1990.
(Informative, simple text accompanies realistic pictures of snakes, Gila monsters, lizards, and other reptiles.)

Spiders

Allen, Judy. *Are You a Spider?* Kingfisher Publishing. New York. 2000.
(With simple text and colorful drawings, introduces the life cycle of a spider, describing how it hatches, develops, spins webs, and feeds.)

Cole, Joanna. *Spider's Lunch: All About Garden Spiders*. Grosset & Dunlap. New York. 1995.
(This book shows how a garden spider catches its food by carefully building a web and then explains how it must wait patiently to catch something.)

Gibbons, Gail. *Spiders*. Holiday House. New York. 1994.
(This book describes the differences in spider habitats, their behaviors, and their webs.)

Glaser, Linda. *Spectacular Spiders*. Millbrook Press. Brookfield, CT. 1998.
(Tour a young girl's yard where she explains how garden spiders trap insects for food in their webs and how they leave silk draglines wherever they go.)

Middle Grades Fiction

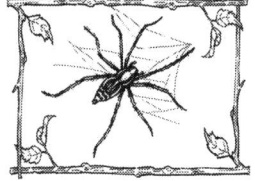

Food Chains and Webs

Reif, Patricia et. al. *The Magic School Bus Gets Eaten: A Book About Food Chains*. Scholastic, Inc. New York. 1996.
(Mrs. Frizzle and her class take a trip to the ocean and try to avoid becoming a part of the food chain.)

Insects

James, Mary. *Shoebag*. Scholastic, Inc. New York. 1992.
(The story of a young cockroach named "Shoebag" who is transformed into a human boy and adopted by a family, but longs to return to his real family and true form.)

Reptiles and Amphibians

Schneider, Rex. *The Wide-Mouthed Frog*. Stemmer House. Baltimore, MD. 1991.
(The wide-mouthed frog decides he is too good to eat flies, so he asks other animals for something better to eat—until he meets the alligator, who likes to eat wide-mouthed frogs.)

Spiders

Arkhurst, Joyce Cooper. *The Adventures of Spider: West African Folktales*. Little Brown and Co. New York. 1992.
(Several classic West African folk tales retold including why spiders live in ceilings and dark corners.)

White, E.B. *Charlotte's Web*. HarperTrophy. New York. 1999.
(The classic story of Wilbur the pig and his spider friend Charlotte, who keeps him from being butchered by writing messages in her web.)

Middle Grades Non-Fiction

Adaptations and Camouflage
Bennett, Paul. *Catching a Meal*. Thomson Learning. New York. 1994.
(Describes how many animals, including chameleons, frogs, ladybugs, and spiders catch their food.)

Bennett, Paul. *Changing Shape*. Thomson Learning. New York. 1994.
(Describes the life cycles of animals that change shape including butterflies, moths, ladybugs, frogs, and toads.)

Powzyk, Joyce. *Animal Camouflage: A Closer Look*. Bradbury Press. New York. 1990.
(Introduces general ways animals camouflage themselves—such as coloration, mimicry, and disguise—and discusses how specific animals protect themselves using these techniques.)

Food Chains and Webs
Kalman, Bobbie, and Jacqueline Languille. *What Are Food Chains and Webs?* Crabtree Publishing Company. New York. 1998.
(Introduces food chains and webs, featuring both herbivores and carnivores. Energy, food production, and decomposition in various ecosystems are also discussed.)

McKinney, Barbara Shaw. *Pass the Energy, Please!* Dawn Publications. Nevada City, CA. 2000.
(Rhyming text and illustrations present nature's food chains—from a simple seed to a top predator—illustrating their natural links.)

Riley, Peter. *Food Chains*. Franklin Watts. New York. 1999.
(Introduces the basic science behind food chains and presents experiments to show how they work.)

Sabin, Francine. *Ecosystems and Food Chains*. Troll Associates. Mahwah, NJ. 1986.
(Explains the natural patterns by which plants and animals depend upon each other and the environment for food, and emphasizes the dangers of pesticides and other human interference with the ecosystem.)

Insects
Berger, Melvin and Gilda. *How Do Flies Walk Upside Down? Questions and Answers About Insects*. Scholastic, Inc. New York. 1999.
(A series of questions and answers provides information about the physical characteristics, senses, eating habits, life cycles, and behavior of different insects.)

Johnson, Sylvia A. *Beetles*. Lerner Publications. Minneapolis, MN. 1982.
(Introduces members of the beetle family, discussing their development, environment, and life cycle.)

Johnson, Sylvia A. *Silkworms*. Lerner Publications. Minneapolis, MN. 1982.
(With exceptional photographs and informative, easy-to-understand text, this book explores the life cycle of the silkworm from egg to larva, to pupa, to adult moth.)

Llewellyn, Claire. *Ladybugs (Keeping Minibeasts)*. Franklin Watts. New York. 2002.
(Discusses the feeding and reproductive habits of ladybugs and suggests ways to raise them and breed them domestically. Includes close-up color photographs.)

Lovett, Sarah. *Extremely Weird Insects*. Avalon Travel Publishing, John Muir Publications. New York. 1996.
(Describes such unusual insects as the net-winged beetle, the brush-snouted weevil, and the peanut-head bug.)

Mason, Adrienne. *Mealworms: Raise Them, Watch Them, See Them Change*. Kids Can Press. Buffalo, NY. 2001.
(Using an attractive, kid-friendly format, this book describes the anatomy and life cycle of a mealworm. It also includes simple experiments and instructions for starting a mealworm farm.)

Mound, Laurence. *Eyewitness Books: Insect*. DK Publishing. New York. 2000.
(A photo essay about insects and their crucial roles in the lives of other living things.)

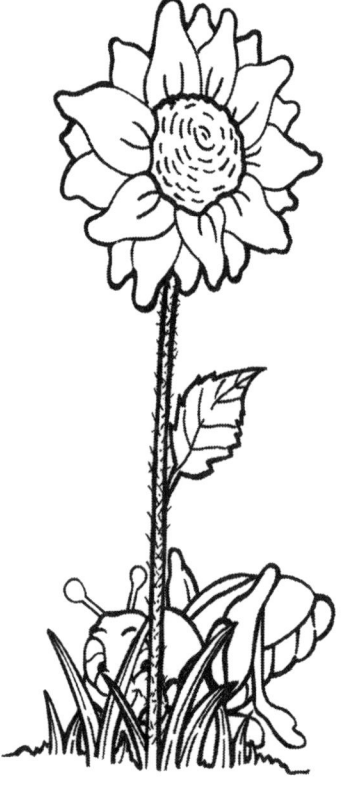

Mudd, Maria M. *The Beetle*. Intervisual Books Inc. Santa Monica, CA. 2001.
(Fully illustrated in color and full of pop-ups, the pages of this unique portfolio discuss in accurate and lively detail the fascinating life cycle of beetles—how they develop from egg to larva to adult beetle—as well as how they fly, mate, and survive.)

Pascoe, Elaine. *Beetles (Nature Close-Up)*. Blackbirch Marketing. Woodbridge, CT. 2001.
(Describes the physical and behavioral characteristics of beetles and offers suggestions for experiments and simple activities.)

Ross, Michael Elsohn. *Ladybugology*. Carolrhoda Books, Inc. Minneapolis, MN. 1998.
(Informative and engaging. Explores questions about ladybugs asked by students and shows how the students answered the questions with experiments.)

Schaffer, Donna. *Mealworms*. Bridgestone Books. Mankato, MN. 1999.
(With detailed close-up photography and easy-to-read text, this book explores each phase in the life cycle of a mealworm.)

Schaffer, Donna. *Silkworms*. Bridgestone Books. Mankato, MN. 1999.
(With detailed close-up photography and easy-to-read text, this book explores each phase in the life cycle of a silkworm.)

Wangberg, James K. *Do Bees Sneeze? And Other Questions Kids Ask About Insects*. Fulcrum Publishing. Golden, CO. 1997.
(Questions and answers explore the insect world, in such categories as body parts and functions, behavior, and habitats. Includes questions and answers about non-insects such as spiders. Also includes projects for students to try.)

Whalley, Paul. *Butterfly & Moth (Eyewitness Books)*. DK Publishing, Inc. New York. 2000.
(Photographs and text explore the behavior and life cycles of butterflies and moths, examining mating rituals, camouflage, habitat, growth from pupa to larva to adult, and other aspects.)

Reptiles and Amphibians

Berger, Melvin and Gilda. *How Do Frogs Swallow With Their Eyes? Questions and Answers About Amphibians*. Scholastic, Inc. New York. 2002.
(Presents a wide variety of information about all amphibians in a question-and-answer format. Beautiful, realistic illustrations accompany easy-to-understand text.)

Clarke, Barry. *Eyewitness Books: Amphibian*. DK Publishing. New York. 2000.
(Stunning real-life photographs accompany informative text about frogs, toads, salamanders, newts, and other amphibians.)

Frogs (Face-to-Face). Scholastic, Inc. New York. 2001.
(Looks at several specific kinds of frogs and describes how frogs eat, how they move, and how they grow from eggs.)

Lovett, Sarah. *Extremely Weird Frogs*. Avalon Travel Publishing, John Muir. New York. 1996.
(Text and stunning up-close photos introduce a variety of unusual and beautiful frogs.)

Lovett, Sarah. *Extremely Weird Reptiles*. Avalon Travel Publishing, John Muir Publications. New York. 1996.
(Describes a variety of reptiles, including Jackson's chameleon, the snake-necked turtle of Australia, and the eyelash pit viper.)

Spiders

Bailey, Jill. *How Spiders Make Their Webs*. Benchmark Books. New York. 1997.
(Describes different kinds of spiders, the types of webs they spin, and the various purposes the webs serve.)

Facklam, Margery. *Spiders and Their Web Sites*. Little, Brown and Company. Boston. 2001.
(Illustrations and text provide a close-up look at the physical characteristics and habits of 12 different spiders and daddy longlegs.)

Lovett, Sarah. *Extremely Weird Spiders*. Avalon Travel Publishing, John Muir Publications. New York. 1996.
(Text and stunning up-close photos introduce unusual spiders.)

Parsons, Alexandra. *Amazing Spiders*. Alfred A. Knopf. New York. 1990.
(Photography and detailed artwork show how a spider spins a web. Read about a spider that flings poisonous hairs at its enemies and spiders as large as dinner plates.)

Ross, Michael Elsohn. *Spiderology*. Carolrhoda Books, Inc. Minneapolis, MN. 2000.
(Informative and engaging. Explores questions about spiders asked by students and shows how the students answered the questions with experiments.)

Schnieper, Claudia. *Amazing Spiders*. Carolrhoda Books, Inc. Minneapolis, MN. 2003.
(Introduces the varieties, appearance, behavior, and life cycle of spiders.)

Other

Johansson, Philip. *The Dry Desert: A Web of Life*. Enslow Publishers, Inc. Berkeley Heights, NJ. 2004.
(Describes the climate, seasons, plants, and animals of the desert. Includes a map and many color photos as well as a glossary and brief lists of recommended books and web sites.)

Johansson, Philip. *The Forested Taiga: A Web of Life*. Enslow Publishers, Inc. Berkeley Heights, NJ. 2004.
(Describes the climate, seasons, plants, and animals of the taiga. Includes a map and many color photos as well as a glossary and brief lists of recommended books and web sites.)

Johansson, Philip. *The Frozen Tundra: A Web of Life*. Enslow Publishers, Inc. Berkeley Heights, NJ. 2004.
(Describes the climate, seasons, plants, and animals of the "frozen desert." Includes a map and many color photos as well as a glossary and brief lists of recommended books and web sites.)

Johansson, Philip. *The Temperate Forest: A Web of Life.* Enslow Publishers, Inc. Berkeley Heights, NJ. 2004.
(Describes the climate, seasons, plants, and animals of the temperate forest. Includes a map and many color photos as well as a glossary and brief lists of recommended books and web sites.)

Johansson, Philip. *The Tropical Rain Forest: A Web of Life.* Enslow Publishers, Inc. Berkeley Heights, NJ. 2004.
(Describes the climate, seasons, plants, and animals of the tropical rain forest. Includes a map and many color photos as well as a glossary and brief lists of recommended books and web sites.)

Johansson, Philip. *The Wide Open Grasslands: A Web of Life.* Enslow Publishers, Inc. Berkeley Heights, NJ. 2004.
(Describes the climate, seasons, plants, and animals of the grasslands. Includes a map and many color photos as well as a glossary and brief lists of recommended books and web sites.)

Kalman, Bobbie. *What is a Biome?* Crabtree Publishing. New York. 1998.
(Introduces biomes, showing and describing the main kinds and discussing their location, climate, and plant and animal life.)

Parker, Steve. *Eyewitness Books: Fish.* DK Publishing. New York. 2000.
(A photo essay about the natural world of fish and their importance in human life.)

Ross, Michael Elsohn. *Rolypolyology.* Carolrhoda Books, Inc. Minneapolis, MN. 1996.
(Informative and engaging. Explores questions about rolypolies asked by students and shows how the students answered the questions with experiments.)

Teacher Resources

Adaptations and Camouflage
Downer, John. *Weird Nature: An Astonishing Exploration of Nature's Strangest Behavior.* Firefly Books. Buffalo, NY. 2002.
(Includes a section about how chameleons eat.)

Forsyth, Adrian. *The Architecture of Animals.* Camden House. Ontario. 1991.
(Explores the homes that various animals make, including spider webs, and examines their impact on the environment.)

Insects
Imes, Rick. *The Practical Entomologist.* Simon & Schuster, Inc. New York. 1992.
(An introductory look at the world of insects, this book examines each order, describes anatomy and life cycle, and gives tips for observing and collecting insects.)

Reptiles and Amphibians
Parsons, Harry. *The Nature of Frogs: Amphibians with Attitude.* GreyStone Books. Berkley, CA. 2000.
(Full of stunning color photographs accompanied by informative text describing each frog's habitat, behavior, and mating habits, among other things.)

Spiders
Levi, Herbert W. *Spiders and Their Kin.* St. Martin's Press. New York. 2003.
(This Golden Guide has information about spiders and their habitats, growth, characteristics, growth, courtship, enemies, and more.)

Rubber Band Books:
Reading in the Content Area

The rubber band books given on the following pages offer valuable content information presented in a kid-friendly way. Each student can be given his or her own book to keep and refer to at a later date. These books also provide a great home link, as students can take them home and share the information they are learning with their parents. To assemble a book, follow these simple instructions:

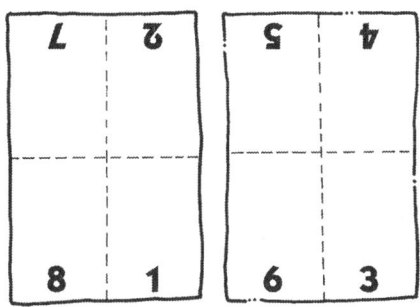 

A #19 rubber band fits perfectly; if these are not available, clip off the inside corners of the book to fit a smaller rubber band.

In several cases, these rubber band books are used in conjunction with a specific activity, but they are also appropriate as introductions to the study of specific critters or concepts.

Silkworms
The Critter Connection

Have you ever felt a piece of soft silk fabric and wondered how it was made? Real silk comes from silkworms, the larval stage of the silkworm moth. A long time ago, silkworms lived in the wild on mulberry trees. Over the years they have been so domesticated that they are unable to survive without the help of humans.

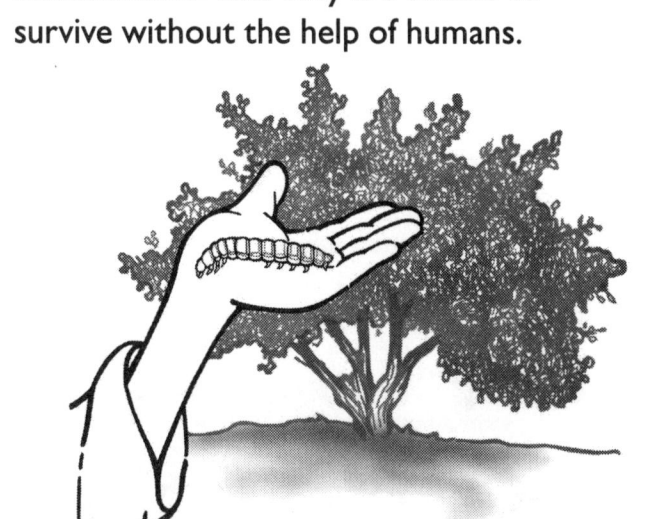

Life Cycle of a Silkworm

Adult silkworm moths cannot fly and do not eat. Their only job is to reproduce themselves by mating and laying more eggs. They may live for up to a week. Females lay between 300 and 500 eggs in neat rows during this time. The eggs are yellow at first, but quickly change to a gray color. If these eggs are saved and refrigerated, a whole new group of silkworms can be hatched the following spring.

Silkworms go through a four-stage life cycle, like many insects. They begin as eggs. When the silkworms first hatch from the eggs, they are less than 3 millimeters long and have large black heads. They begin to eat as soon as they hatch and grow very quickly.

Are you a picky eater? Silkworms sure are. The only food a silkworm will eat is mulberry leaves. As it eats, a silkworm quickly grows too large for its skin. When this happens, it has to molt, or shed its old skin. A silkworm molts four times during the larval stage. By the time it is ready to spin its cocoon, a silkworm is about 10,000 times bigger than when it was hatched.

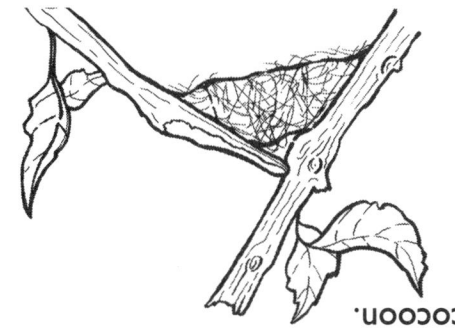

When a silkworm is ready to spin a cocoon, it will stop eating and move its head back and forth repeatedly. It will find a corner or other place to attach its cocoon and begin to spin. A silkworm's cocoon is made of silk, which comes from its spinneret, located just below its mouth. It moves its head back and forth in a figure-eight pattern, forming the walls of the cocoon. After two to four days, the silkworm will be completely enclosed inside its cocoon.

Safe inside its cocoon, a silkworm molts one more time, becoming a pupa. During this stage, the final metamorphosis from silkworm to moth gradually takes place. After about three weeks, the metamorphosis is complete, and the adult moth is ready to break free from the cocoon. It does this by dissolving the silk and pushing its body through the hole.

Many insects go through four stages of growth. The changes that occur in their life cycles are called metamorphoses. Metamorphosis is a Greek word that means to transform or change. In a complete metamorphosis, an insect goes through four stages.

In the wild, mealworms are usually found in dark, damp places. They are scavengers who like to eat rotting grain and cereal. They will eat grain, cereal, flour, bran, bread, crackers, meat scraps, feathers, and the bodies of dead insects.

The Critter Connection
Mealworms

A mealworm is not a worm at all. It is the larva of the yellow mealworm beetle, which is a kind of darkling beetle. A mealworm begins its life as a white, bean-shaped egg about one millimeter long. The eggs usually take about one week to hatch.

After hatching, the larvae (mealworms) begin to eat. They eat grains, bran, and cereals, and soon become too large for their hard skins. A mealworm sheds its skin several times before it begins the pupa stage. This is called molting. The larval stage of the mealworm beetle lasts anywhere from several weeks to two years, depending on the conditions.

At the end of the larval stage, the mealworm encloses itself in a pupa case, and transforms into the adult beetle. The pupa stage lasts between one and three weeks.

When the adult comes out of the pupa case, it is white. It gradually turns brown and finally black. The adult beetle has wings, but does not usually fly. It can hop about 10 to 12 centimeters. The adult beetle lives only a few months. The female may lay up to 500 eggs before she dies, and the life cycle starts all over again.

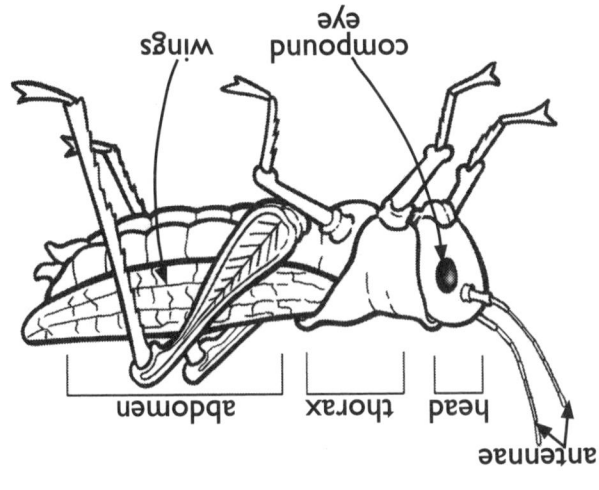

Body Parts of a Typical Insect

Labels: wings, compound eye, antennae, head, thorax, abdomen

Most insects go through a four-stage life cycle called *metamorphosis*. The four stages of metamorphosis are egg, larva, pupa, and adult. Some insects that go through metamorphosis are flies, moths, beetles, and butterflies.

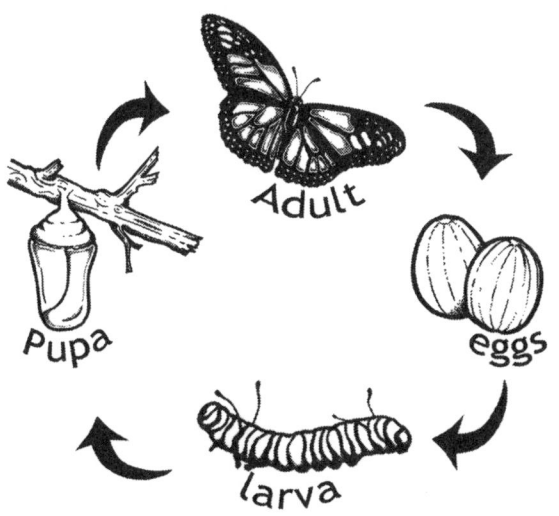

Adult, eggs, larva, Pupa

Insects

The Critter Connection

Insects have been on Earth for more than 300 million years. There are more insects than all other kinds of animals put together. Entomologists (people who study insects) have probably discovered and identified less than half of the insect species that exist on Earth.

Insects are arthropods belonging to the class *Insecta.* The word arthropod means "jointed legs." Other arthropods—insects' closest relatives—are spiders, crabs, lobsters, centipedes, and millipedes.

Crab

Spider

Lobster

The major role of many insects is pollination—a very necessary service to the plants and animals of the Earth. Insects also eat other insects and serve as food for many larger animals. Not all insects are good for humans, however. Some—such as mosquitoes, fleas, and ticks—carry disease, and others—like locusts and grasshoppers—can destroy crops.

Disease

How can you tell if a critter is an insect?

Must have:
- ☑ Three body parts: the head; the thorax, or chest; and the abdomen, or hind body
- ☑ Six legs
- ☑ A hard exterior skeleton, or exoskeleton

Most have:
- ☐ Wings
- ☐ Two compound eyes, or a number of small, simple eyes
- ☐ Antennae for smelling

If your critter fits these descriptions, you have found an insect.

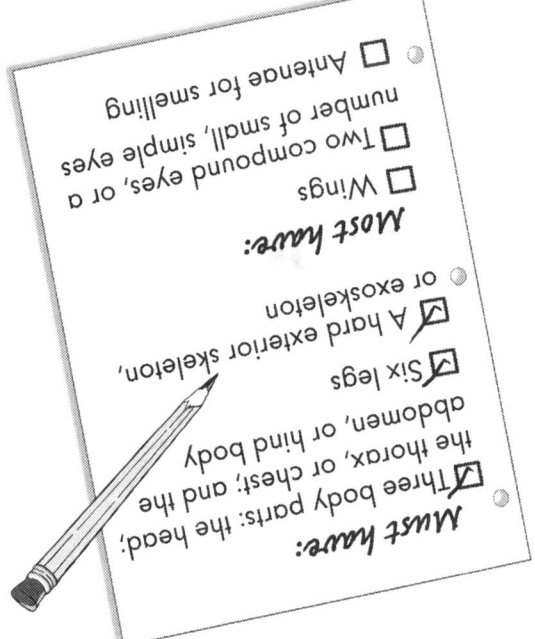

There are four major groups of insects:

BEETLES

MOSQUITOES AND GNATS

FLIES,

WASPS, BEES AND ANTS

BUTTERFLIES AND MOTHS

Spiders
The Critter Connection

Spider silk is useful to humans as well as spiders. It has been used by primitive people for fishing nets, lures, bags, and headdresses. The silk has also been used to make the cross hairs in astronomical telescopes, levels, and surveying equipment.

Spiders have been on Earth for hundreds of millions of years. More than 30,000 different species of spiders are known, over 2000 of which live in the United States. If you want to get away from spiders you'll have to travel to Antarctica—they live on every other continent in the world!

Spiders are not insects, but they do belong to the same phylum—*Arthropoda.* Their class is *Arachnida,* and their order is *Araneae.*

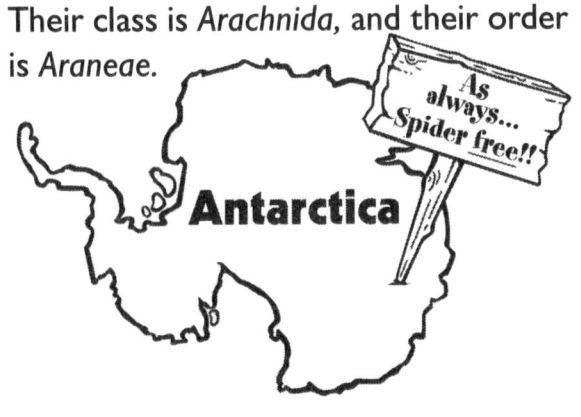

As always... Spider free!!

Antarctica

Spiders produce silk from glands in their bodies. As the liquid shoots out of the six spinnerets and is exposed to the air, it hardens. The spinnerets control and shape the silk which can be thin, thick, dry, sticky, soft, stretchy, or beaded. Spiders use their silk for webs, traps, nests, wrappings, and as draglines. Females wrap their eggs with silk. Baby spiders release silk and use it to carry themselves in the wind. This is called ballooning.

The word *Arachnida* comes from a Greek myth. According to this myth, Arachne lived in Greece many centuries ago. Her spinning and weaving were so skilled that she challenged Athena, the goddess of weaving and handicrafts, to a contest. Her tapestry was so perfect that, the story says, Athena became enraged and turned her into a spider, condemning her to perpetual spinning. The English word spider comes from the German word spinner, meaning one who spins.

Spiders have an exoskeleton like insects, but have eight legs instead of six. The spider's body has two parts. The head and the thorax (cephalothorax) are one part, the abdomen is the other. Spiders have no wings, antennae, or feelers, and usually have eight simple eyes. When a spider grows, it sheds its outer skin (molts) instead of going through a metamorphosis. Most spiders live about one year, but some species, such as tarantulas, may live as long as 20 years!

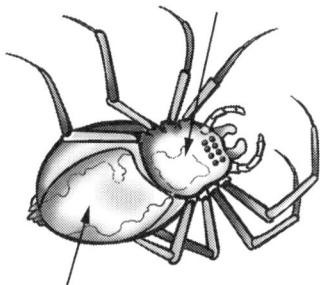

cephalothorax

abdomen

The major role of spiders is to eat insects. Every year spiders eat millions and millions of insects that destroy grain crops and eat green leaves. Most spiders are timid, harmless, and quite helpful. The two North American spiders to watch out for are the black widow and the brown recluse, which are harmful to humans.

BEWARE! **Brown recluse**

BEWARE! **Black widow**

A spider does not chew its food. It paralyzes or kills its prey with its fangs, and then uses a digestive juice to turn the bug's tissues into a liquid that it sucks into its stomach.

The Critter Connection
Frogs and Toads

Frogs and toads are amazing creatures. They can be found on every continent except Antarctica. Some live on the ground, some live in water, and some even live in trees. They come in almost every color of the rainbow from bright red, to orange, green, or blue. The smallest are less than one centimeter long, while the largest are almost one foot long. Scientists have classified more than 4100 species of frogs and toads.

Frogs and toads are amphibians. An amphibian is an animal that lives part of its life in the water and part of its life out of the water. Some other amphibians are newts and salamanders. Frogs and toads are unique amphibians because they have four legs and no tails.

Frogs and toads need water to reproduce. Female frogs and toads lay eggs in water (lake, pond, puddle), and male frogs and toads fertilize the eggs. When the eggs hatch they are tadpoles. Tadpoles will change a lot before they become adult frogs or toads. This process is called metamorphosis.

It's easy to see the difference between a frog and a salamander, but what about between a frog and a toad? There are some ways to tell, but be careful! There are always exceptions to the rule. In fact, scientists often use "frog" as a generic term to talk about both frogs and toads.

To be 100% sure if an animal is a toad or a frog, you have to know the scientific name. (So-called "true frogs" are in the family *Ranidae*. "True toads" are in the family *Bufonidae*.) But if you spot a four-legged friend in the wild, you can use these lists to be about 95% sure.

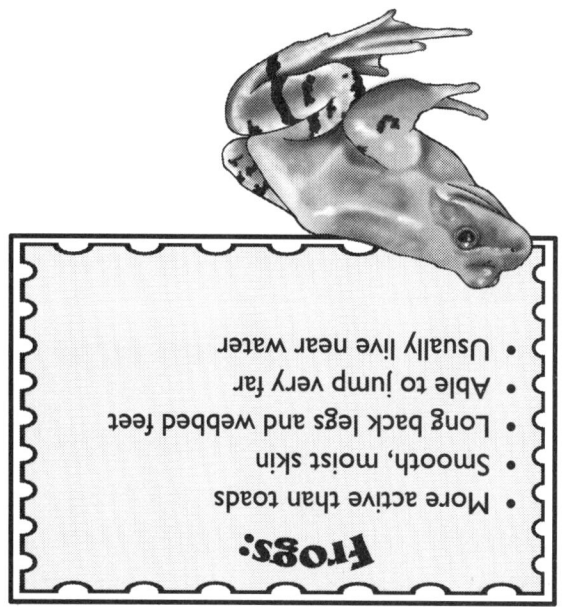

Frogs:
- More active than toads
- Smooth, moist skin
- Long back legs and webbed feet
- Able to jump very far
- Usually live near water

Toads:
- Less active than frogs
- Dry, warty skin
- Short, stubby legs, feet more like claws
- Walk, waddle, or make short hops
- Don't have to be near water

Even though they have differences, there are many things that are the same about frogs and toads. They both have large eyes on the tops of their heads. This lets them see all around to help find food and stay out of danger. All frogs and toads begin their lives as tadpoles living in the water. The males of many species of frogs and toads croak or sing to the females.

Camouflage

The Critter Connection

Can you think of another type of camouflage that was not defined here? Describe it below.

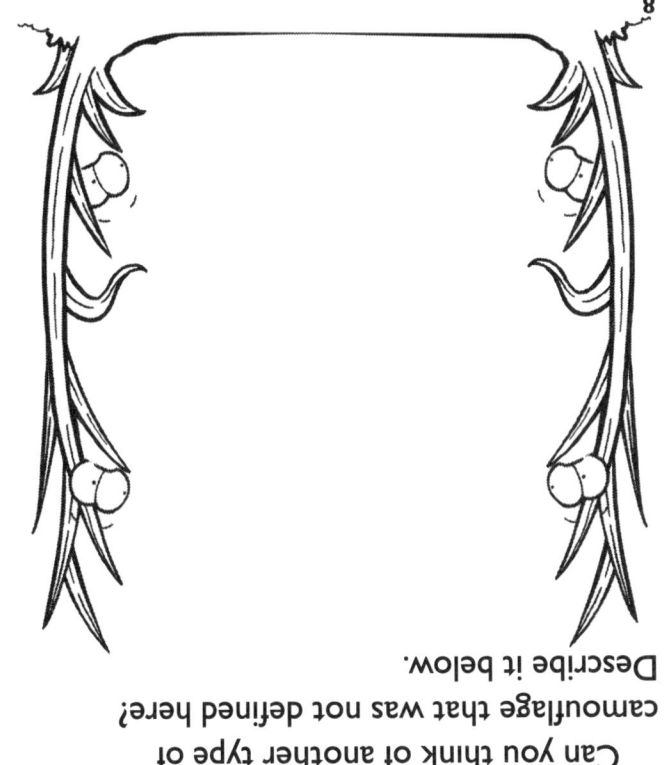

Camouflage is any use of shape, pattern, and/or color that helps an animal to be less visible. Animals use many different types of camouflage for many different reasons. Some use camouflage to hide from predators, while others use it to hide while hunting prey.

There are other kinds of camouflage that animals use, but the four just described are the most common. Think of at least one more example of each kind of camouflage and write it below.

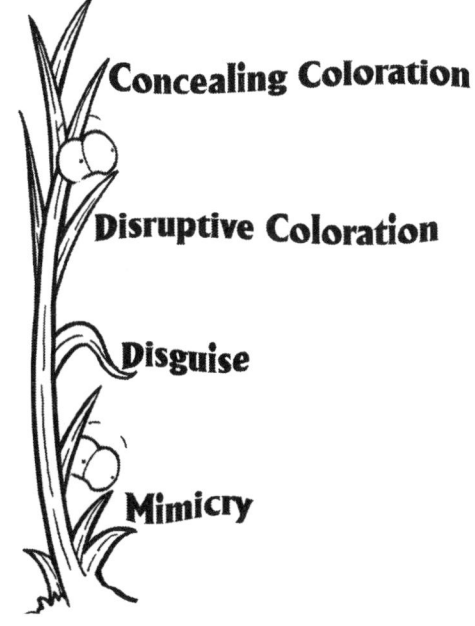

Concealing Coloration

Disruptive Coloration

Disguise

Mimicry

Concealing coloration is a type of camouflage in which animals have body colors that match their surroundings. Animals such as the bobwhite and baby deer have colorings that blend into their normal surroundings. If they change locations, the camouflage no longer works. Others, like chameleons and octopuses, are able to change the color of their bodies to match their surroundings.

Disruptive coloration can be seen in animals like the leopard. Its spots help to disrupt its shape and make it difficult to tell where the leopard ends and its surroundings begin. In order for this kind of camouflage to work, the animal must stay completely still—any movement will give it away.

Mimicry is another form of camouflage in which one kind of animal looks very much like another animal. The king snake looks very much like the poisonous coral snake, that most animals avoid. The viceroy butterfly has markings that are almost identical to the monarch butterfly, which birds don't like to eat because of its bad taste. Both of these animals benefit from looking like something they are not.

When an animal uses *disguise* as a form of camouflage, it looks like something else. Walking sticks and leaf insects are two examples of animals that are very well disguised just by the shapes of their bodies. There is another insect called an orchid mantis that looks just like one of the flowers on which it sits. It waits for an unsuspecting moth or bee to come to the flowers and then attacks.

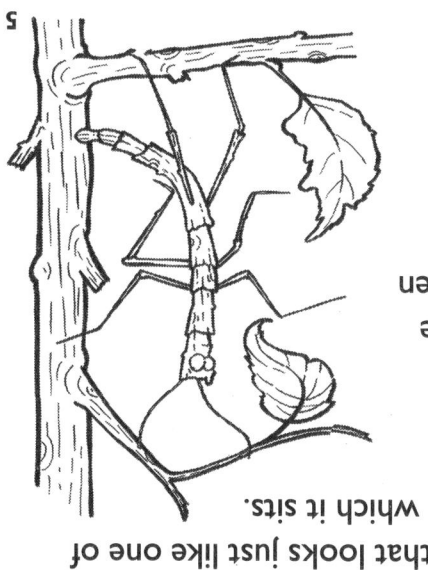

Living things need food to give them energy. A food chain is the path by which energy passes from one living thing to another. All energy originally comes from the sun. Green plants use energy from the sun to make food. Because of this, they are called *producers*.

Here is a sample food web. Look at how many food chains are in the food web.

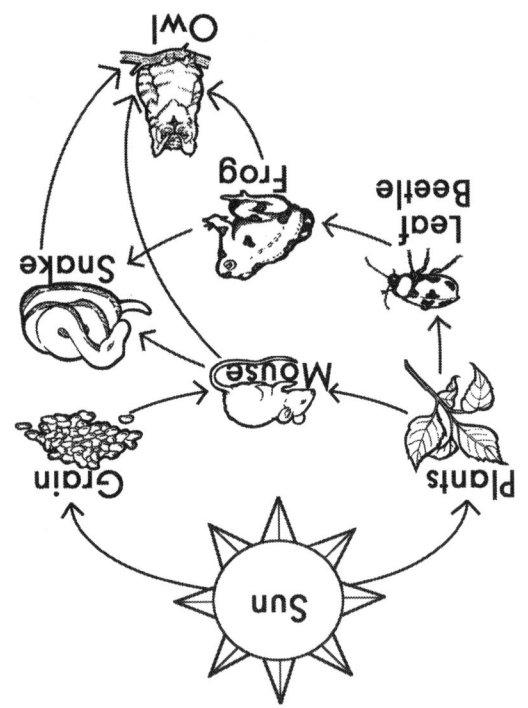

Next time you sit down for a meal, think about where your food is coming from. Are you eating producers, consumers, or both? What kinds of things did the consumers you are eating eat?

Draw a food chain or web that shows something you have eaten recently.

Animals cannot make energy from the sun. They must eat plants and/or other animals to get their energy. They are the *consumers* in a food chain. Consumers that eat only plants are called *herbivores*. Those that eat only meat are called *carnivores,* and those that eat both plants and meat are called *omnivores.*

Herbivore Carnivore Omnivore

Organisms that break down large molecules into smaller parts are called *decomposers.* Fungi (like mushrooms, mold, and yeast) and bacteria are two kinds of decomposers. Decomposers are an important link in any food chain or web. They return the nutrients that are in a living thing to the soil. Without decomposers, future generations of plants would not have the nutrients they need to grow.

A food chain is a simplified way to look at the energy that passes from producers to consumers. One food chain might look something like this:

Sun => Grass => Cow => Human

A food web is a more realistic way of looking at the relationships between plants and animals in an environment. A food web is created when several food chains are linked together. Predators eat a variety of prey. It is likely that a predator from one food chain would be linked to the prey of a different chain.

Lenses & Ladybugs

Ladybug, ladybug, fly away home,
Your house is on fire and your children are alone...

Some people think this is a warning to ladybugs because in September farmers used to burn the hop fields, which were full of ladybugs.

Topic
Observing ladybugs

Key Question
What does a ladybug look like through a hand lens and microscope?

Learning Goal
Students will use a hand lens and a microscope to observe features of a ladybug.

Guiding Documents
Project 2061 Benchmarks
- A lot can be learned about plants and animals by observing them closely, but care must be taken to know the needs of living things and how to provide for them in the classroom.
- Magnifiers help people see things they could not see without them.
- Animals eat plants or other animals for food and may also use plants (or even other animals) for shelter and nesting.
- Draw pictures that correctly portray at least some features of the thing being described.

NRC Standards
- Employ simple equipment and tools to gather data and extend the senses.
- Simple instruments, such as magnifiers, thermometers, and rulers provide more information than scientists obtain using only their senses.

Math
Graphing

Science
Life science
   insect anatomy
      ladybugs

Integrated Processes
Observing
Recording
Comparing and contrasting

Materials
For the class:
   ladybug habitat (see *Management*)
   a supply of ladybugs
   microscope
   raisins
   floor graph

For each student:
   one ladybug
   portion cup with lid (see *Management 6*)
   raisins
   hand lens
   science journal
   two 6" paper plates
   one piece of 6" x 6" tissue paper
   one paper fastener
   black sticky dots or construction paper dots
   2 paper flower stamens (from craft store, see
      Management 9)
   three chenille stems (see *Management 9*)
   scissors
   glue
   crayons or paint

Background Information
Ladybird beetles, commonly called ladybugs, are found in most parts of the United States. They are usually located in neglected garden areas where their main food supply, the aphid, is busy feeding on plants. This common garden insect makes a wonderful vehicle for students to investigate the anatomy of an insect. It also serves as a great specimen to use when teaching the use of science tools such as hand lenses and microscopes.

Ladybugs are quite diverse in their appearance. They vary in color, but are usually bright red, orange, or yellow with black spots. This coloration serves as a warning to their predators that they have an unpleasant taste. The location and number of spots also varies. One species alone has been found to have anywhere from two to six spots, and has also been found completely black. Students will

CRITTERS   © 2004 AIMS Education Foundation

enjoy comparing the spot patterns of their ladybugs with the spot patterns of others.

While observing the ladybug through a hand lens and then a microscope, students will discover the insect's three major parts: head, thorax, and abdomen. They will also be able to observe the placement of the legs, wings (hard outer wings that serve as a covering to protect the transparent inner wings used for flight), eyes, and antennae.

Management

1. If ladybugs are not available in your area, you can order live specimens from one of the companies listed in the *Resources*.
2. If you are collecting ladybugs from your area, be sure to include leaves from the plants where you find them. These leaves may have aphids and aphid eggs on them. Raisins can also be used for ladybug food.
3. A simple-to-make critter cage can be fashioned for observing and storing the ladybugs in the classroom. To do this, stand a sheet of fine wire mesh in and around the perimeter of a drip pan from a potted plant container. For example, use a 3-foot tall piece of wire mesh inside two 20" round plastic drip pans. Glue the edges of the wire mesh with a hot glue gun or weave nylon fishing line between the two edges. To provide access to the habitat, do not permanently attach the top pan to the wire mesh. Pour prepared plaster of Paris into the bottom pan and insert and secure wire mesh. Let dry. Be sure to include a food source of leaves with aphids and/or raisins. To provide water for the ladybugs, simply sprinkle a few drops at the bottom of the container each day.

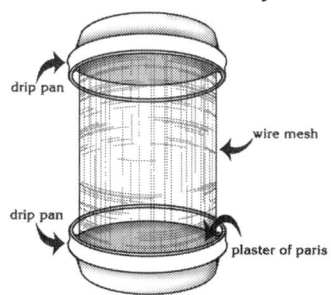

4. This activity assumes that students have been introduced to the proper use of a hand lens and microscope.
5. This activity is divided into three parts: *Part One* utilizes observations made with hand lenses; *Part Two* allows students to look at ladybugs with a microscope; *Part Three* has students make paper plate ladybug models for graphing purposes.
6. For making observations, place one ladybug for each student in a small, transparent container with a lid that will fit under a microscope. The small portion cups found at restaurant supply stores, or small clear plastic boxes will work.
7. Place a raisin in the container with each ladybug.
8. Prepare a large class graph that will accommodate the paper plate ladybug models the students will be making.
9. The chenille stems will be used for the ladybug's legs in *Part Three*. If students are able, let them determine how to make six legs from the three chenille stems. If they are not developmentally ready for this, precut the legs before doing *Part Three*. The flower stamens are used to represent the antennae. If they are not available, use another chenille stem.
10. It may be easier for the students if the tissue paper wings are precut. To do this, trace around the paper plate and cut out the circles. Fold the circles in half. Cut up this center fold to within 2 cm of the upper edge.

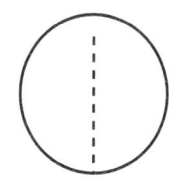

Procedure

Part One

1. Direct the students to draw a picture in their science journals of what they think a ladybug looks like.
2. Allow the students to share these drawings with each other and discuss similarities and differences.
3. Tell the students that they are going to compare their drawings with actual ladybugs. Introduce the ladybug collection.
4. Distribute a small container which houses one ladybug and a raisin to each student.
5. Tell them that you are going to give them a science tool that will help them observe their ladybugs. Distribute the hand lenses. Tell the students to be very careful with their living creatures. After some observation time, discuss how the ladybugs appear larger through the hand lens.
6. Direct them to draw a second picture of a ladybug, incorporating what they observed.
7. Have students compare and contrast their pictures with others and with their own previous drawings made without the hand lenses.
8. Allow the students to share what they saw. Make a list on a large wall chart as to all the things the students observed.
9. If the students saw things that they couldn't identify (like the aphids) or simply wanted to know more, take time to discuss these things. This may be a good time to stop and to send someone or a group of students to the library to conduct some research. Depending on the grade level of your students, you may want to assign this research for homework.

Part Two
1. Bring out the microscopes.
2. Direct the students to use the same ladybug they used with the hand lenses. Distribute the microscopes or set up a station where students can take their ladybugs to the microscopes for observation.
3. Allow the students to rediscover that, as with the hand lens, a microscope makes the ladybug appear larger than it really is. Suggest to the students that they will be able to see many more interesting things about the ladybug that they could not see with just their eyes or the hand lens. Allow the students a period of free exploration followed by a time to again draw their ladybugs with a closer eye on the subject!
4. Continue the discussion of how their drawings are changing with the use of the science tools of a hand lens and microscope.

Part Three
1. Distribute the ladybug construction materials to each student. Allow students to paint or color the plates to look like their ladybugs. Use the black sticky dots or the black construction paper cut into circles to represent the dots. Have students attach the chenille stems in the appropriate places on the underside of the model as legs. The flower stamens can be used as antennae.
2. Direct the students to use the tissue paper for the transparent wings which fit under the hard red outer wings (the second paper plate cut partially in half). Attach the top paper plate and tissue paper wings to the bottom paper plates using a paper fastener.
3. Once their ladybug models have been constructed, direct the students to use them to respond to a class floor graph. A suggested question for the graph is *How Many Spots Does Your Ladybug Have?*

Connecting Learning
1. Do all ladybugs have the same pattern of spot markings?
2. Do all ladybugs have the same number of spots? How many did your ladybug have?
3. How many wings does your ladybug have?
4. Describe the way it flies. [in a straight line, up and down, darting, flapping]
5. Where do you think your ladybug would live in nature?
6. Is the head small or large? Which way does it move?
7. Does it have eyes? Where are they? How many eyes does it have?
8. Does your ladybug have antennae? What do you think these are used for?
9. How many legs does it have? Are they jointed like yours?
10. What color and shape is the body? How many body parts are there? Does the body change shape when it moves? Is it symmetrical?
11. How did your illustrations change when you used a magnifying lens and microscope to look at your ladybug?
12. Describe things about the ladybug you noticed with the hand lens/microscope that you did not notice when you just used your eyes.
13. Using our class graph, what can you tell me about the ladybugs in our classroom?
14. Using our class graph, what do you think you might know about most ladybugs?
15. What do you still wonder about ladybugs? Where do you think you could find the answers to your questions?

Extensions
1. Take the students outside to an observation area. Direct them to use their hand lenses to observe places where they think a ladybug might be. Encourage the students to imagine that they are the size of a ladybug and to discuss what the area would be like.
2. Explain to students that not all people like to live in the same types of places. Some may dislike hot weather; other people might not like the rainy weather; some may prefer to live alone, while others like to live in families or with friends. Tell students that they are going to do a habitat survey to find out where ladybugs are located. Have them use a page in their journal to list different kinds of places where they can look. (Start with these: in the air, in long grass, in short grass, in trees or bushes, under rocks, in water, on plants, in dead leaves, in soil, in dead wood, on the sidewalk.) Direct the students to take a hand lens, pencil, clipboard, and journal on a habitat-survey walk. Tell the students to put a check next to the place they have listed every time they find a ladybug. Try doing this survey more than once; on a warm, sunny day, and then on a cloudy, damp day and see if you get the same or different results. Be sure to visit the same places both days. **Warning: When you do your habitat-survey walk, remember that if you lift a rock or piece of wood, you may be lifting the roof of some critter's home. Remember to gently put it back!**
3. Have students graph the results from their habitat survey. Suggested questions for the graph are *Where Did You Find Your Ladybug? Where Did You Find the Most Ladybugs?*
4. Use other types of critters such as a praying mantis, mealworms, or earthworms.

5. Place flowers in the ladybug habitat. While the ladybugs are searching for aphids, they will crawl around the stamen and will gather pollen onto their bodies. Allow the students to place a ladybug under the hand lens or microscope to see the amount of pollen gathered. When the ladybug flies to another flower, some of this pollen is rubbed off onto the stigma, thus pollination occurs.
6. Maintain the critter habitats over a long period of time giving the students the responsibility for the care of the living organism.

Resources
If ladybugs are not available in your area, you can order live specimens from one of many companies that sell them. Two such companies are listed here:

Insect Lore
1-800-LIVE BUG
http://www.insectlore.com

Carolina Biological Supply Company
1-800-334-5551
http://www.carolina.com/

Curriculum Correlation
Allen, Judy. *Are You a Ladybug?* Kingfisher. New York. 2000.

Brown, Ruth. *Ladybug, Ladybug*. Puffin. New York. 1992.

Carle, Eric. *The Grouchy Ladybug*. Scholastic, Inc. New York. 1977.

Home Link
Send the science journals home so the students can continue their ladybug-survey walk there. Ask the students to share their findings with the class.

What I see

What I see

What I see

CRITTERS © 2004 AIMS Education Foundation

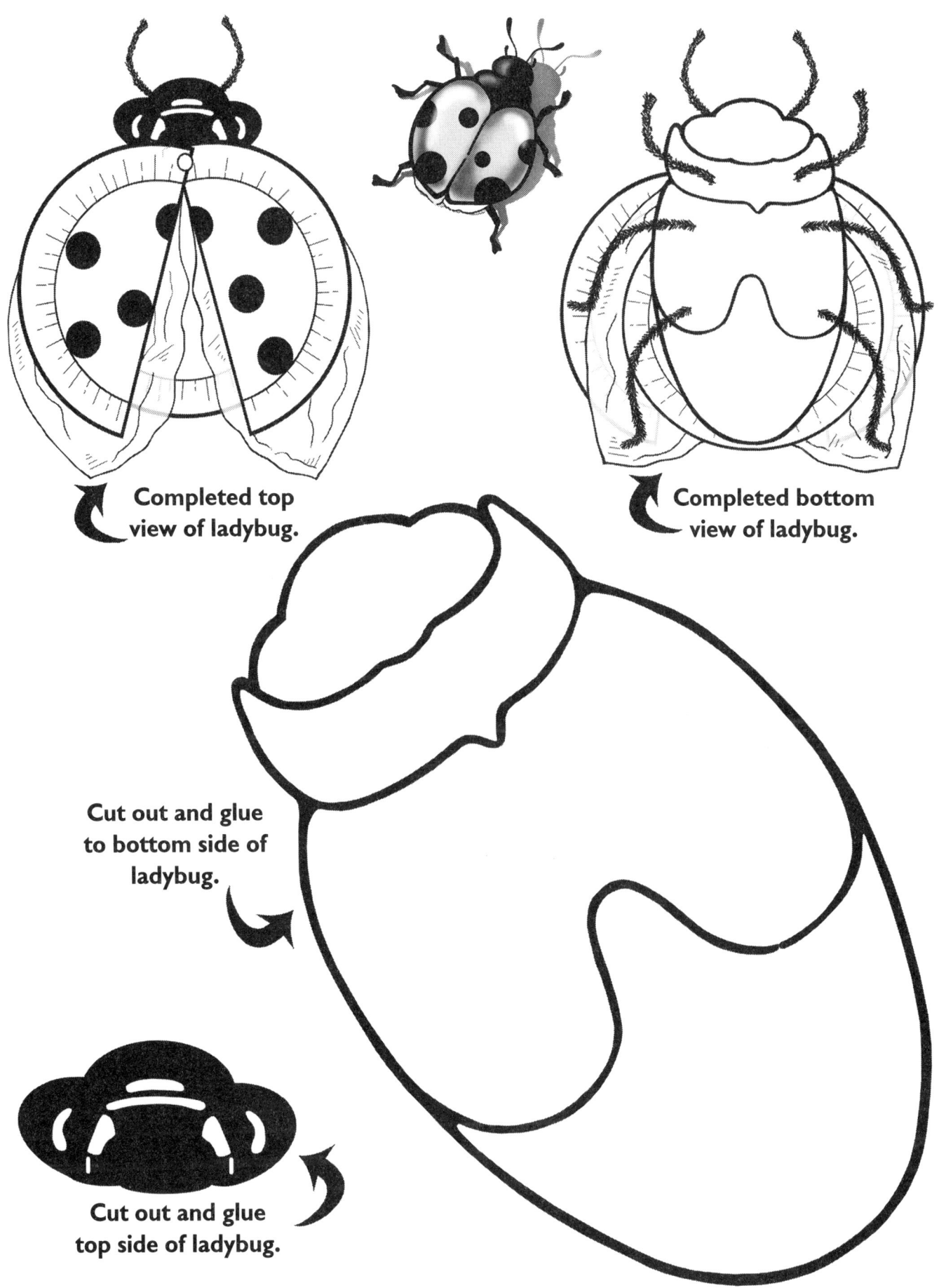

Spider Spoofs and Proofs

Topic
Spiders

Key Questions
1. How are spiders in children's stories different from real spiders?
2. How does one type of spider and its web compare to another type of spider?

Learning Goals
Students will:
1. listen to and read fictional stories about spiders, and
2. observe real spiders in their natural habitats.

Guiding Documents
Project 2061 Benchmarks
- *Stories sometimes give plants and animals attributes they really do not have.*
- *There is variation among individuals of one kind within a population.*
- *Animals eat plants or other animals for food and may also use plants (or even other animals) for shelter and nesting.*

NRC Standard
- *Organisms have basic needs. For example, animals need air, water, and food; plants require air, water, nutrients, and light. Organisms can survive only in environments, and distinct environments support the life of different types of organisms.*

Science
Life science
   animals
      spiders

Integrated Processes
Observing
Recording
Comparing and contrasting
Communicating

Materials
For each student:
   journal (see *Management 2*)
   clipboard (see *Management 4*)

For the class:
   sentence strips
   study prints (see *Management 5*)

Background Information
Many children's literature books portray spiders as animals having attributes very different than real spiders. Through the use of fiction, some books have spiders talking, playing musical instruments, dancing, and more. Without an opportunity to compare and contrast the differences between real spiders and fictional ones, the young learner may construct many misconceptions about spiders.

This activity asks students to use their observational skills to compare and contrast a fictional spider with a real spider. Through this comparison, the young learner will also discover many attributes of spiders and information about their habitats and behaviors. An important part of this activity is the sharing of information among students: what they see, what they think, and what it makes them wonder about. Young learners should have time to record and to talk about what they observe and to compare their observations with those of others.

Through observing different spiders, the young learners will discover that there are many different species of spiders and differences among spiders of the same species. These differences will be evident not only through physical markings and body structures, but also in behaviors among the spiders the children will observe.

Through observations of the spiders in their natural habitats, the students see how plants and other structures in the surroundings are used by the spiders for shelter and as supports for their webs.

Management
1. Discuss safety rules to follow when observing spiders.
   - Do not touch any of the spiders.
   - Do not reach inside a dark place.
   - Ask an adult to turn over rocks and wood.
   - Do not destroy the habitat of the spider.
2. Duplicate one journal per student. Copy the journal pages front to back so that students have a single page with three spaces to record. For more recording space, copy the second page front to back a second time, giving students seven spaces in which to record.

CRITTERS © 2004 AIMS Education Foundation

3. Gather several fiction and non-fiction books about spiders. See *Curriculum Correlation* for a list of suggested books.
4. Prepare a field scientist clipboard as illustrated at the end of this teacher text.
5. Study prints of four different spiders have been included with this activity. Spiders were chosen that are found throughout the United States and Canada. If none of these spiders is found on your schoolyard, you may want to check with a local entomologist for pictures of common spiders in your area.
6. The Cellar Spider included in this activity *(Pholcus phalangioides)* is commonly called a "daddy longlegs" in many parts of the country. It should not be confused with the real daddy longlegs, which is not a spider, but an arthropod.
7. Optional: Copy the spider study prints on one side of the paper and the spider facts on the other side to provide ready information about each spider.

Procedure
Part One
1. Read a series of fictional books about spiders to the class.
2. Ask the students to describe the behaviors and attributes of the spiders in the stories. List the students' responses on sentence strips and place them in a pocket chart or on the bulletin board.
3. Read a series of non-fictional books about spiders to the class.
4. Ask the students to describe the behaviors and attributes of the spiders in these books. List the students' responses on sentence strips and place them in a pocket chart or on the bulletin board.
5. Direct the students to compare and contrast the listed attributes. Ask them to arrange the sentence strips into two groups. *Things that are* **true** *about real spiders* and *Things that are* **not true** *about real spiders*. Ask the students to explain how they decided which statements were true and which were not.
6. Continue adding to these groups as you read additional fiction and non-fiction stories and books to the class throughout this study of spiders.

Part Two
1. Give the students a field scientist's clipboard and journal. Lead the students on a walk around the schoolyard. Challenge them to locate evidence of spiders such as new and abandoned webs. Caution the children not to touch the webs and to be on the lookout for the actual spider. Review safety rules about protecting themselves from spider or insect bites. Describe the rules of respect for the habitats of other living creatures as well.

2. Have the students record the various places they locate the webs. Encourage them to describe and/or illustrate the webs. Ask students to record the time of day that they make these observations.
3. Once the students return to the classroom, discuss the information gathered. Discuss the type of environments where they located the spiders. Were they in the dark, in the open, under leaves, on the walls, in the corners of buildings, etc.? Record these observations on sentence strips to be added to the *Things that are true about real spiders* chart.
4. Take the class out for several other observation periods, at different times of the day, for at least three days. Direct them to continue recording their observations and the time of day, and add these observations to the class chart of statements.

Part Three
1. After several days of observation and discussion in their groups, lead the students in a discussion as to why they think the spiders are living in different types of environments. What are the spider's needs and how does the environment help the spider satisfy those needs?
2. Show the class a study print of a spider common to the area. Discuss whether or not they saw this type of spider on their walk. If they did see this spider, ask them to describe where they observed it and whether or not it had a web.
3. Ask the students to describe all the different spiders they located. Discuss the similarities and differences among spiders of the same species as well as those among spiders of different species.
4. Read another fictional story about spiders. Now that they have observed the real spiders, have them compare the information about the fictional spider character in the story to the spiders that they have observed on their schoolyard. Have them add to the *Things that are not true about real spiders* group of statements.
5. Review the different true and not true statements they have listed about the fictional spider characters and those they have discovered through their observations and research. Discuss their discoveries.

Connecting Learning
1. Describe where you found evidence of spiders.
2. Was the spider always in the web? Where do you think the spider goes when it isn't in its web? Why do you think it leaves its web?
3. What does a real spider house look like? Is there a dining room table, a couch, a bed, and refrigerator? Why do you think these things are not in a real spider's house?

4. Do all spider webs look the same? Describe how they are different and how they are alike.
5. Do all spiders look the same? Explain.
6. Describe how the spiders use the area and plants in their environments.
7. What were the spiders doing each time you found them?
8. Compare the behavior of the spiders described in the stories we read to the behavior of the real spiders you observed. Explain the differences. Were any of the real spiders dancing, playing an instrument, or talking?
9. Tell what you know about spiders. What do you still wonder about?

Extensions
1. For further study of spiders, follow this activity with *Spying On Spiders*.
2. Ask the students to explore the Internet and libraries for additional information on spiders to share with the class.

Curriculum Correlation
Allen, Judy. *Are You a Spider?* Kingfisher. New York. 2000.

Carle, Eric. *The Very Busy Spider*. Putnam Publishing. New York. 1995.

Cole, Joanna. *Spider's Lunch: All About Garden Spiders*. Grosset & Dunlap. New York. 1995.

Gibbons, Gail. *Spiders*. Holiday House. New York. 1994.

Glaser, Linda. *Spectacular Spiders*. Millbrook Press. Brookfield, CT. 1998.

Internet Connections
Additional information about spiders can be found at these sites:

Answers to 20 common questions about spiders and information on 10 types of spider families.
http://www.explorit.org/science/spider.html

Watch a quick-time movie of the actual moves made by a spider while constructing a web.
http://www.conservation.unibas.ch/team/zschokke/webconstruction.html

Pages offering links to other spider websites
http://www.amonline.net.au/spiders/resources/links.htm

http://arthur.k12.il.us/arthurgs/spidlink.htm

Field Scientist Clipboards
These clipboards can be taken on walking or riding field trips. Students or helpers can use them as sturdy supports to record experiences, reactions, or comments made during the explorations.
- Cut a 9" x 12" piece of heavy cardboard.
- Cover the front and back of the cardboard with adhesive-backed paper. Seal the edges.
- To add more support, using masking tape to frame the edges all around.
- Attach two jumbo paper clips to the top of the clipboard for holding an 8.5" x 11" piece of paper.
- Cut a 24" length of string, cord, or yarn. Tie one end around a sharpened pencil. Secure the string to the pencil with masking tape. Knot the other end of the string and attach it to the back of the clipboard with masking tape.

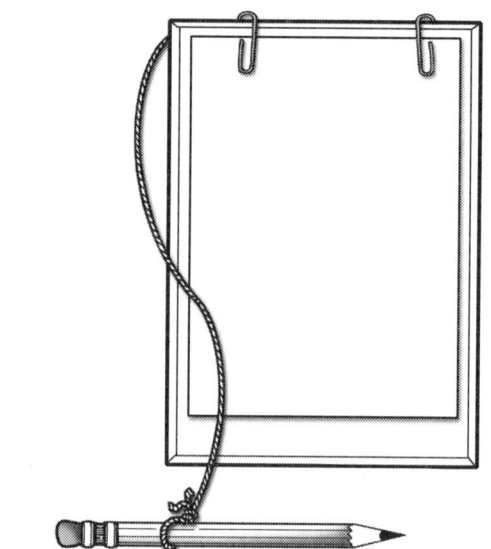

Spider Spoofs & Proofs

The heroic story of a girl and her tuffet.

Ms. Muffet

This book belongs to _____

M. Muffet

Spider Spoofs and Proofs

My Spider...

Time _____

CRITTERS

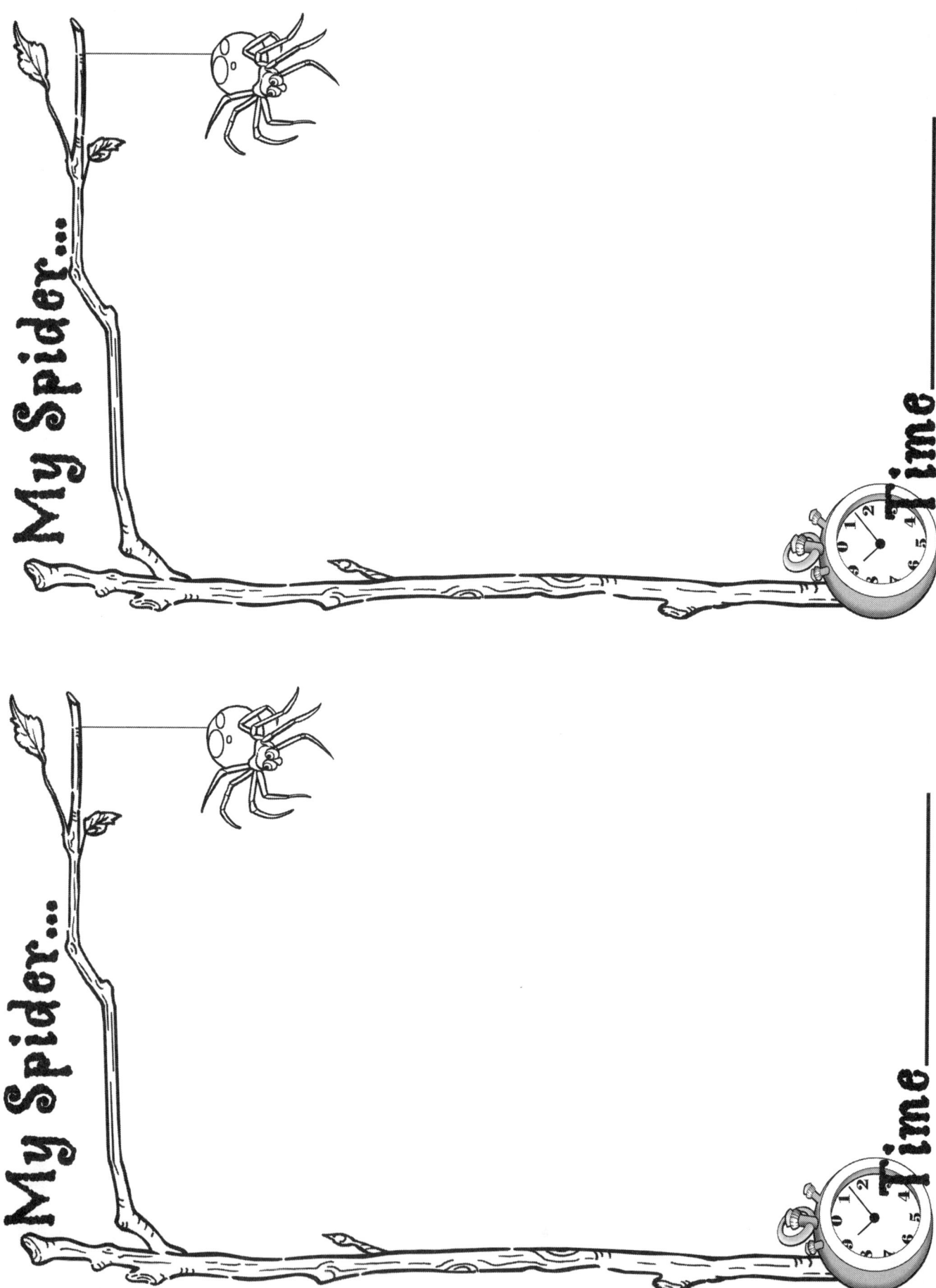

Comb-Footed Spiders (Family: Theridiidae)
American House Spider
Size: 4-6 mm, female is larger than male
Color: Yellowish brown with black and gray streaks and splotches on sides. Male has orange legs, female has yellow legs with black bands.
Silk: Spins an irregular, sticky web beneath ceilings or in window frames in houses, barns, etc.; uses the silk to wrap its prey; forms a brownish egg case.
Food: Insects
Other Information: Have comblike bristles on the hind legs that are used for wrapping their prey in silk.

Daddy-longlegs (Family: Pholcidae)
Cellar Spider
Size: 2-8 mm
Color: Whitish or gray. Abdomen has dark and light areas forming a pattern. Knee segments are darker than the rest of the legs.
Silk: Loose, untidy webs, built under the eaves and in corners of buildings and cellars; This spider does not use its web to catch prey; the web is where it "hangs out." It wraps its prey in silk to immobilize it, then the spider proceeds to eat the prey. The female carries the round silken egg sac in her jaws.
Food: Insects, spiders (even its own kind), and decaying organic matter
Other Information: This spider is also known as the vibrating spider because when it is disturbed it vibrates its web to scare off the enemy. The venom of this spider is extremely poisonous; however, it is not considered dangerous to humans because its fangs are too small to penetrate the skin.

Funnel Web Weavers (Family: Agelenidae)
Grass Spider
Size: 15-20 mm, female is larger than male
Color: Yellowish with a center stripe and two dark and light bands on edges. Abdomen has a light stripe down the middle with gray, darker bands on the edge.
Silk: Spins a funnel-shaped web of non-sticky silk in shrubs, tall grasses, on stone fences, and on the walls of buildings. Forms an egg sac which is kept outside the web.
Food: Insects
Other Information: The grass spider hides in the bottom of the web's funnel. When an insect hits the web, the spider speeds out, bites the insect, and pulls it into the funnel.

Orb Weavers (Family: Araneidae)
Marbled Orb Weaver
Size: 6-19 mm, female is larger than male
Color: Reddish yellow to yellow. Dark lines run along sides and middle with yellow spots near the lines. Abdomen looks marbled with browns and purples.
Silk: Spins a spiraling orb web, built on shrubs and tall grasses of meadows; forms white cocoonlike egg sac around orange colored eggs
Food: Insects
Other Information: This spider senses the vibrations of the web if an insect is caught.

Spying on Spiders

Topic
Spiders

Key Questions
How does one type of spider and its web compare to another type of spider?

Learning Goal
Students will observe real spiders in captivity.

Guiding Documents
Project 2061 Benchmarks
- *There is variation among individuals of one kind within a population.*
- *Animals eat plants or other animals for food and may also use plants (or even other animals) for shelter and nesting.*

NRC Standard
- *Organisms have basic needs. For example, animals need air, water, and food; plants require air, water, nutrients, and light. Organisms can survive only in environments in which their needs can be met, and distinct environments support the life of different types of organisms.*

Science
Life science
    animals
        spiders

Integrated Processes
Observing
Collecting and recording data
Comparing and contrasting
Communicating

Materials
For each student:
    clipboard (see *Management 4*)
    journal (see *Management 5*)

For the class:
    spider habitats (see *Management 2*)
    two or three different types of spiders
    hand lenses
    chart paper

Background Information
This activity asks students to use their observational skills to carefully observe spiders in order to gather information about them and their webs. They will compare and contrast spiders of different species as well as those within the same species. Through this comparison, the young learner will discover many attributes of spiders and information about their habitats and behaviors. An important part of the students' explorations is telling others what they see, what they think, and what it makes them wonder about. Young learners should have time to record and talk about what they observe and to compare their observations with those of others.

Through observing different spiders in their classroom, the young learner will discover that there are differences among spiders of the same species as well as among spiders of many different species. These differences will be evident not only through physical markings and body structures, but also in the behaviors the children will observe.

Through observations of the spiders in a close-up, controlled habitat, the students will see how plants and other structures in the surroundings are used by the spiders.

Sample Observations of a Cellar Spider
- At first you can't see the web. You know something is there because the spider is walking back and forth in the middle of the twigs.
- Some parts of the web have more silk than other parts.
- Some insects stick to the web, but others don't.
- The spider changes and fixes its web a lot.
- The water seems to keep the spider on the twig. The spider went down to the water, but then it went back up to the top of the twig.
- The spider is usually at the top of the twig.
- When the spider first got on the twig it was very busy. Now that it has a web, it's not so busy anymore.
- When the web gets touched, the spider sometimes moves up and down like it's bouncing.
- The spider can walk all over the web. It doesn't get stuck.
- When the spider eats a bug there is a see-through body left in the web.

CRITTERS   © 2004 AIMS Education Foundation

- When a bug was in the web, the spider didn't go to it right away. After a couple of hours, it ate the bug.
- The spider has markings on its legs and body.

Management
1. Discuss safety rules to follow when gathering spiders. The students should be allowed to help find the spiders; however, the teacher needs to do the actual catching of the spiders. It is best to begin the activity with only one or two spiders. Add more if additional habitats are constructed. Go over the rules for safety.
   - Do not touch any of the spiders.
   - Do not reach inside a dark place.
   - Ask an adult to turn over rocks and wood.
   - Do not destroy the habitat of the spider.
2. Build a spider environment using a three-inch deep dish (the drip saucer for a plant pot works well). Secure a 12-18 inch forked twig from a tree or shrub in the center of the container as follows: Place the twig into the small end of a Unifix cube. Push the cube into a ball of clay that is not soluble in water. Flatten the clay and place it, the cube, and the twig in the center of the container. Build up small rocks around the cube to help stabilize the twig. Fill the container with water to cover the rocks.

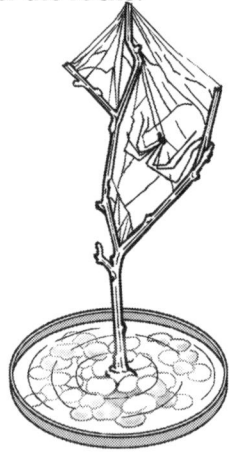

3. Before doing this activity, do the activity *Spider Spoofs and Proofs* in which students listen to and read stories (fiction and nonfiction) about spiders and observe real spiders in their natural habitats.
4. Have students use the field scientist clipboards from *Spider Spoofs and Proofs*.
5. If possible, copy the *Spying on Spiders* journal onto a single page, front and back. Each student will need one journal for up to three days of observation. If you have students observe the spiders for more than three days, they will need multiple copies of the interior pages of the journal.

6. To feed the spiders while in captivity, either catch small insects or purchase pinhead crickets at the local bait shop or pet supply store. Place the insects on the twig or in the web. It is not recommended that you keep the spiders in captivity for more than three or four days. Please be sure to release them in the area where you captured them.

Procedure
Part One
1. Review the information gathered by the students in the activity *Spider Spoofs and Proofs*.
2. Ask the students to describe how they think a spider spins a web. After a bit of discussion, tell them that they are going to try to watch a spider actually spin a web.
3. Display the spider environment. Establish class rules regarding the use of the environment. Explain that you will be placing a spider onto the twig in hopes that it will spin a web. Tell the class that it will be their job to carefully observe the spider and to record and be ready to describe what they see over a period of a few days.
4. Ask the students what they think the purpose of the water around the twig is. [to keep the spider on the twig so that it does not escape into the classroom] Discuss how a web is something very special and that they are not to touch the work of the spider in any way.
5. Place the spider on the twig and invite small groups of students to the observation area. Record their observations on large chart paper. Suggest that they draw the spider, the twig, and anything else they observe. Provide a set of hand lenses for the students to take a closer look at the spider and its web.
6. At different times during the day, schedule each group of students for an observation period at the spider environment. Have them illustrate their observations and record the time of day they make their observations.
7. Continue this observation schedule for two or three days.
8. On the last day, ask the groups to choose a reporter. Line up the reporters in order of the time of day they made their observations. Display the illustrations of each group and allow them to describe their observations.
9. Discuss the changes in the spider's activity as well as the progress of the web. Ask the students to describe everything they noticed about the spider and the spider environment.

10. Once the class is finished with their observations, return the spider to the same place it was originally found.

Part Two
1. Repeat *Part One* using another spider of the same kind as the first. Compare similarities and differences in the behavior and the web that is built by both spiders.
2. Repeat once more using a different type of spider. For example: If you introduced a cellar spider the first time, try using a garden spider—one that spins an orb web—for this round of observations.

Conclusion
1. Gather the students together in front of the sentence strip charts from the previous activity, *Spider Spoofs and Proofs*. Review the different *true* and *not true statements* they have listed about the fictional spider characters and those they have discovered through their own observations and research.

Connecting Learning
1. Do all spider webs look the same? Describe how they are different and how they are alike.
2. Do all spiders look the same? Explain.
3. What did you notice about the spiders you observed in their natural environment compared to the spiders you observed in the classroom environment?
4. Compare the behavior of the spiders described in the stories we read to the behavior of the real spiders you observed. Explain the differences.
5. Describe how the spiders use the area and plants in their environment.
6. Describe the process of the construction of the webs. Use your illustrations and notes to tell about the changes that took place during the observation period of several days.
7. If the spider ate something, explain when it ate (during the day or night). Describe what they left behind and anything else you observed.
9. Tell about what you know about spiders. What do you still wonder about?

Extension
Ask the students to explore the Internet and libraries for additional information on spiders to share with the class.

Curriculum Correlation
Allen, Judy. *Are You a Spider?* Kingfisher. New York. 2000.

Carle, Eric. *The Very Busy Spider*. Putnam Publishing. New York. 1995.

Cole, Joanna. *Spider's Lunch: All About Garden Spiders*. Grosset & Dunlap. New York. 1995.

Gibbons, Gail. *Spiders*. Holiday House. New York. 1994.

Glaser, Linda. *Spectacular Spiders*. Millbrook Press. Brookfield, CT. 1998.

Internet Connections
Additional information about spiders can be found at these sites:

Answers to 20 common questions about spiders and information on 10 types of spider families:
http://www.explorit.org/science/spider.html

Watch a quick-time movie of the actual moves made by a orb-weaving spider while constructing a web:
http://www.conservation.unibas.ch/team/zschokke/webconstruction.html

Great lists of links to other spider sites on the Internet:
http://www.amonline.net.au/spiders/resources/links.htm

http://arthur.k12.il.us/arthurgs/spidlink.htm

Spying On Spiders
Journal

Name _____

Spider Environment
Observations

Date: _____ Time: _____

What I observed:

How it looked:

CRITTERS © 2004 AIMS Education Foundation

Spider Environment Observations

Date: Time:

What I observed:

How it looked:

Spider Environment Observations

Date: Time:

What I observed:

How it looked:

CRITTERS © 2004 AIMS Education Foundation

Growing Pains

Topic
Silkworms

Key Question
How will the number of leaves a silkworm eats affect how much it grows?

Learning Goal
Students will determine if the amount of food a silkworm gets affects its growth rate.

Guiding Documents
Project 2061 Benchmarks
• Most living things need water, food, and air.
• Animals eat plants or other animals for food and may also use plants (or even other animals) for shelter and nesting.

NRC Standards
• Organisms have basic needs. For example, animals need air, water, and food; plants require air, water, nutrients, and light. Organisms can survive only in environments, and distinct environments support the life of different types of organisms.
• All animals depend on plants. Some animals eat plants for food. Other animals eat animals that eat the plants.
• Employ simple equipment and tools to gather data and extend the senses.
• Use data to construct a reasonable explanation.

NCTM Standards 2000*
• Represent data using tables and graphs such as line plots, bar graphs, and line graphs
• Use tools to measure

Math
Measurement
  linear
Data analysis
  line graph
  averages

Science
Life science
  animals
    silkworms

Integrated Processes
Observing
Predicting
Collecting and recording data
Comparing and contrasting
Drawing conclusions

Materials
Silkworms, one per student
Containers for silkworms (see *Management 3*)
Fresh mulberry leaves (see *Management 1*)
Hand lenses, one per student
Metric rulers (see *Management 4*)
Chart paper
Student pages
The Critter Connection: Silkworms
#19 rubber bands

Background Information
Silkworms are the larval stage of the silkworm moth. They are easy to care for, and their life cycle is short enough to make them ideal insects for classroom observation. Although silkworms used to live in the wild, over the years they have become so domesticated that they are now incapable of survival without human assistance. Their diet consists entirely of mulberry leaves, which they consume in great quantities daily. Much of the silk fabric and thread we have in the United States comes from silkworms that are raised commercially on farms.

Management
1. Be sure there is a mulberry tree nearby or that you have access to a daily supply of fresh mulberry leaves! Silkworms have a voracious appetite and will not eat any other kind of leaves.
2. This activity is more easily done if you begin about 10 days after silkworms emerge. They are very difficult to measure prior to this time. You will need to collect data for at least two weeks.
3. Silkworms are not very mobile. They can be safely housed in containers about four inches deep. For the individual student containers, small shoeboxes work well. If you intend to keep the silkworms through their entire life cycle, you will need to provide egg cartons or twigs in the corners of the containers for the silkworms to spin their cocoons.

CRITTERS 34 © 2004 AIMS Education Foundation

4. Be sure your rulers are precise to the nearest millimeter.
5. Silkworm eggs may be ordered through biological supply companies and from a variety of sources on the Internet. See *Resources* for details.

Procedure
Day One
1. Distribute the pages for the silkworm rubber band book and assist students with assembly.
2. Read the book together as a class and discuss the information it contains.
3. Distribute one silkworm in a small container and a hand lens to each student. Tell students that it is safe to handle the silkworms, but remind them to be gentle.
4. Allow time for observations and record students' discoveries on a piece of chart paper.
5. Have students complete the observation page and discuss their findings.

Day Two
1. Discuss the *Key Question*.
2. As a class, brainstorm ways to find out how the amount of food affects the growth of a silkworm.
3. Assign a special diet of one, two, three, four, or five leaves per day to each student's silkworm. (Try to have equal numbers of silkworms for each diet.)
4. Hand out the second student page and explain that each student will be responsible for feeding his or her silkworm the proper number of leaves and measuring its length each school day.
5. Allow time over the next 14 school days for students to collect and record their data.

Conclusion
1. After all of the data are collected, distribute the final two student pages. Have students whose silkworms had the same number of leaves each day share their data to find the average final length for that particular diet.
2. Come together as a class and have each group share their data so that everyone can complete the line graph on the last page.
3. Close with a final time of class discussion and sharing where students analyze the results of their experiment.

Connecting Learning
1. Did the silkworms that were fed the most grow the longest? Is this what you predicted? Why or why not?
2. What was the average length for silkworms eating each number of leaves?
3. What did your line graph tell you about the effect the amount of food eaten has on the growth of silkworms?
4. Using your data, is it possible to predict the continued growth of a silkworm with a certain diet? What would your prediction be?
5. What are you wondering now?

Extensions
1. Try to find a pattern of growth relative to the number of leaves given.
2. Find out if there are other *Lepidoptera* (moths or butterflies) that cannot fly.
3. Compare silkworms with mealworms and butterfly larvae.
4. Study the life cycle of the silkworm by continuing the observations begun in this activity. The life cycle can be drawn on the "Metamorphosis Wheel" found in the *Mealworms on Stage* activity. A life cycle chart can also be made using a grain of rice to represent the egg, a short length of white pipe cleaner to represent the silkworm larva, a lump of cotton or an actual cocoon to represent the pupal stage, and a picture or drawing of the silkworm moth to represent the adult stage.
5. Make a silkworm book to chronicle the life of your silkworm.

Curriculum Correlation
Literature
Hong, Lily Toy. *The Empress and the Silkworm.* Albert Whitman & Co. Morton Grove, IL. 1995.

Johnson, Sylvia A. *Silkworms.* Lerner Publications. Minneapolis, MN. 1982.

Schaffer, Donna. *Silkworms.* Bridgestone Books. Mankato, MN. 1999.

Art
Collect pieces of silk fabrics and make a display.

Resources
Live silkworm eggs are available from the following suppliers:

Insect Lore
1-800-LIVE BUG
http://www.insectlore.com

Carolina Biological Supply Company
1-800-334-5551
http://www.carolina.com/

Aurora Silk
503-286-4149
http://www.aurorasilk.com/shop/eggs.shtml

* Reprinted with permission from *Principles and Standards for School Mathematics*, 2000 by the National Council of Teachers of Mathematics. All rights reserved.

GROWING PAINS

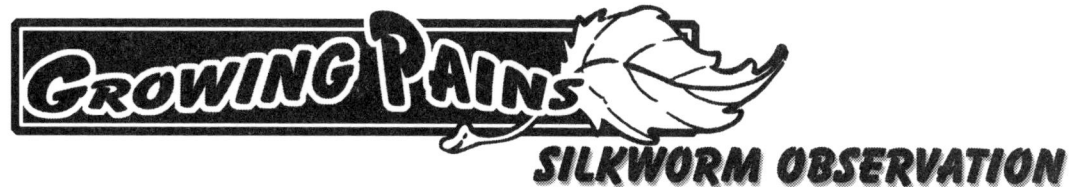

SILKWORM OBSERVATION

Study your silkworm carefully. Color the picture that looks the most like a real silkworm.

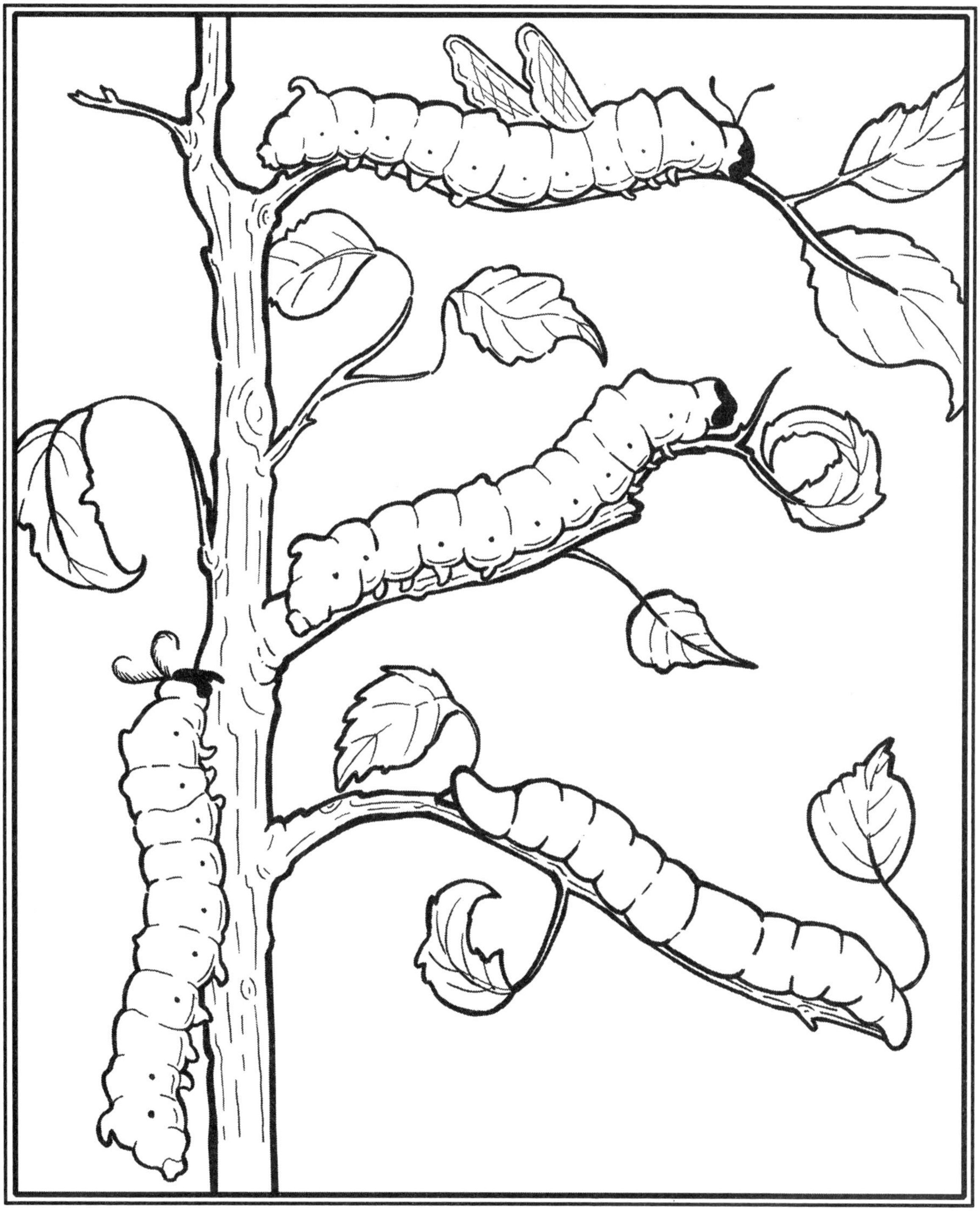

Watch your silkworm eat. What do you observe?

Does the amount of food available affect the growth of a silkworm?

How will the number of leaves you feed your silkworm affect its growth? Write your prediction here.

Color in the number of leaves that you will be feeding your silkworm each day.

Measure your silkworm each day. Record your measurements in the table below.

Record your observations about your silkworm and how it is growing. Describe any interesting or surprising discoveries.

My Silkworm

Day	Length
1	mm
2	mm
3	mm
4	mm
5	mm
6	mm
7	mm
8	mm
9	mm
10	mm
11	mm
12	mm
13	mm
14	mm

Record the final length for each silkworm that was fed the same number of leaves each day. Find the average length by adding all of the lengths together and dividing by the number of silkworms.

Silkworm lengths for _____ leaves

Dietician	Final Length
1	mm
2	mm
3	mm
4	mm
5	mm
6	mm
7	mm
Average	mm

Were the final lengths of the silkworms similar or different? What might account for some of the differences?

Which number of leaves resulted in the longest silkworms? Does this surprise you? Why or why not?

Record the average length for every number of leaves in the table below.

Number of leaves	Average length
1	mm
2	mm
3	mm
4	mm
5	mm

CRITTERS

Use the data from the table on the previous page to make a line graph. Be sure to write in the appropriate measurements along the y-axis.

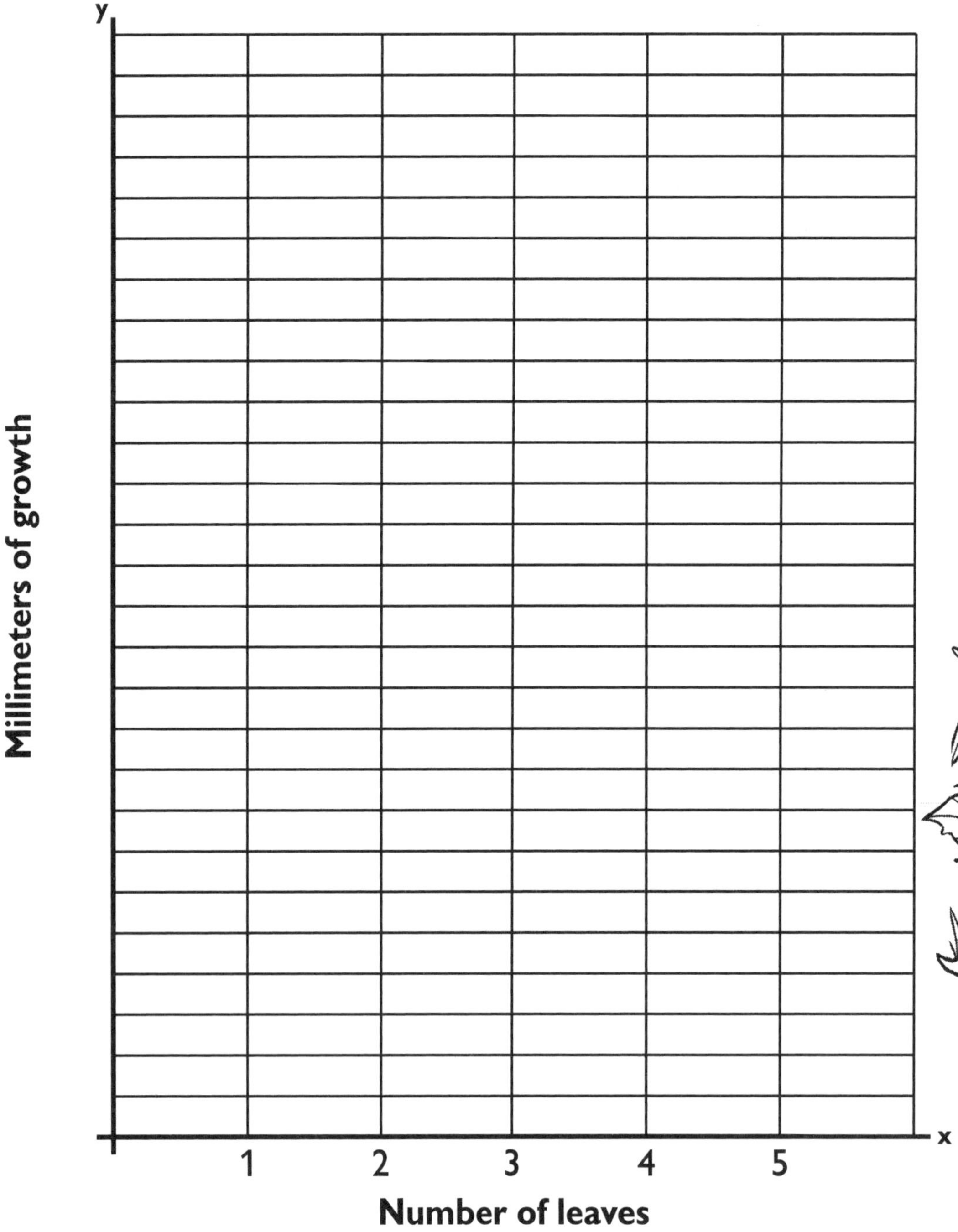

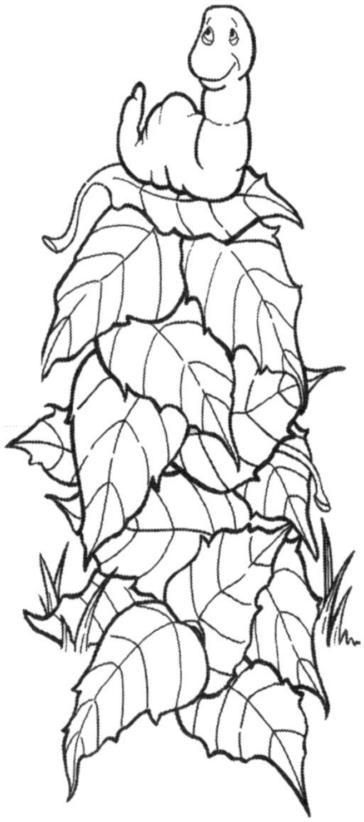

Millimeters of growth (y-axis)

Number of leaves (x-axis): 1, 2, 3, 4, 5

What does the graph tell you about how the number of leaves relates to silkworm growth?

Mealworm Moments, Part One

Topic
Animal behavior

Key Question
What can you learn about mealworms by observing them closely?

Learning Goal
Students will learn about animal behavior through observations of mealworms.

Guiding Documents
Project 2061 Benchmarks
- A lot can be learned about plants and animals by observing them closely, but care must be taken to know the needs of living things and how to provide for them in the classroom.
- Things move in many different ways, such as straight, zigzag, round and round, back and forth, and fast and slow.

NRC Standard
- Employ simple equipment and tools to gather data and extend the senses.

NCTM Standard 2000*
- Use tools to measure

Math
Measurement
　length

Science
Life science
　animal behavior
　　mealworms

Integrated Processes
Observing
Comparing and contrasting
Collecting and recording data
Communicating

Materials
For each student:
　hand lens
　index card
　eyedropper and water
　small transparent plastic bag
　2 mealworms
　white paper, minimum 8.5" x 11"
　My Mealworm Journal

For the class:
　colored pencils
　plastic container (see *Management 1*)
　peat moss (see *Management 1*)
　pieces of apple, zucchini, carrot, or potato
　　(see *Management 2*)
　Parts of a Mealworm page (see *Management 8*)

Background Information
　This activity provides opportunities to observe animal behavior and is designed around the scientific method of investigation. This activity begins with guided exploration of a mealworm and its behavior. It should be followed with *Mealworm Moments, Part Two*, which will continue the investigations in a task card format.
　Mealworms are the larvae of Tenebrid or darkling beetles. Two varieties of mealworms are commonly available as pet food: the common yellow mealworm (*Tenebrio molitor*) and a larger relative called a King Mealworm or Superworm (*Zophobas morio*). A third type, marketed as Giant Mealworms, is actually the same species as the common yellow mealworm, but has been treated with growth hormones and is larger in size. King mealworms are recommended for these investigations, as the smaller and less active varieties are more difficult for young learners to observe.
　King mealworms, like many other insects, have four life stages: egg, larva, pupa, and adult. The adults are black or dark brown and cannot fly. The longest part of their life cycle is spent in the larval stage. In the wild, King mealworms live in tropical climates in locations such as bat caves and the rain forest floor. They are raised commercially as food for many varieties of reptiles, amphibians, small rodents, fish, birds, and other animals.

Management
1. When the King mealworms are not being used in an activity, keep them in a plastic container with sides at least 10 cm above a bed of damp peat moss that is three to five centimeters deep. They prefer temperatures of 70° to 80° Fahrenheit, and should never be kept below 60°.

2. Always keep several pieces of apple, zucchini, potato, or carrot in the container to provide food and moisture. Replace every two to three days to avoid mold.
3. Mealworms can be purchased from a pet store, bait shop, or other supplier. (see *Resources*.)
4. Encourage the students to handle the mealworms gently and with respect. Show the students how to use an index card to pick up the mealworm. Assure those who are reluctant that mealworms cannot bite. Any scratching sensation they feel comes from the mealworms' claws, which are not big enough to damage human skin.

5. Placing the mealworm in a small plastic bag will help keep the mealworm motionless enough for close observation. Demonstrate how to use their thumb and pointer finger to corral the mealworm in the corner of the bag in order to slow it down for observation.

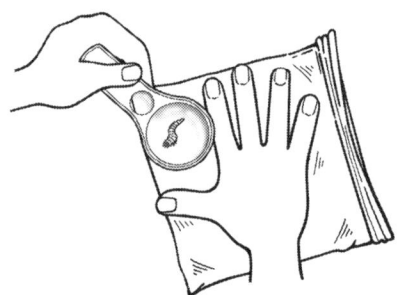

6. It is normal for a healthy mealworm to be inactive at times. If it is dead, it will become stiff and turn dark brown or black.
7. Be aware that mealworms are a potential pest to agriculture and should never be released into the wild. When you are finished with your mealworms, share them with another class or donate them to your local zoo or pet store.
8. Enlarge or make a transparency for the overhead of the *Parts of a Mealworm* to show to the entire class.
9. If possible, copy the *My Mealworm Journal* onto two sheets of paper, front to back. Fold the papers in half and nest them inside each other so that the pages are in order. Staple along the spine to hold the pages together. Each student will need one journal.

Procedure
1. Provide each student with a mealworm in a small transparent plastic bag. Give students time to observe the mealworms and notice everything they can about them. Ask the students to talk about what the mealworms look like, how they feel, and how they behave.
2. Record their responses on chart paper and discuss the observations. Direct the students to observe their mealworms once more to look for characteristics noted by their classmates that they may not have noticed before.
3. Distribute colored pencils and one *My Mealworm Journal* to each student.
4. Ask students to draw their mealworms as accurately as they can, much larger than life. Direct them to show both the color and the shape of the body, head, and legs.
5. Ask the students if they think that all mealworms look alike. Put a second mealworm in each student's bag. Discuss similarities and differences.
6. Guide the discussion to questions about the number of body segments, legs, and antennae that each mealworm has. Have students count these features on each of their mealworms, comparing the information they found. Show them where to record this information in their books. Inform students that critters with six jointed legs are insects. Ask them if mealworms are insects. [yes]
7. Distribute eyedroppers to students and ask them to complete the observations listed on page 4 of their journals.
8. Provide hand lenses or other magnifiers and encourage the students to take a very close look at their mealworms. Ask them to record some of the details they see on page 5 of their journals.
9. Question whether or not all the mealworms are the same length. Direct the students to measure one of their mealworms. Have them record their values in their journals. Allow time for them to generate lists of things longer and shorter than their mealworms.
10. Use the *Parts of a Mealworm* illustration to point out the different parts of a mealworm. Explain that mealworms are not finished growing and that they will look very different when they are adults. Refer to caterpillars becoming butterflies or moths, and ask if anyone can guess what mealworms become. [beetles]
11. Instruct each student to turn one mealworm loose on a piece of paper and observe its behavior. Tell students to be sure that the mealworm stays on the piece of paper. Ask the students to talk about what they notice about the mealworm. Continue recording on the class chart, listing

what they have observed and adding what they would like to find out.
12. In their *My Mealworm Journals*, have the students reflect and record what they learned about mealworms.

Connecting Learning
1. Describe your mealworm. In what ways is it different from the others? In what ways is it the same?
2. Describe the different parts of the mealworm.
3. Why do you think a mealworm needs claws?
4. How did you determine the length of your mealworm? What did you find out about the lengths of our mealworms?
5. Describe the way a mealworm moves (straight, zigzag, round and round, back and forth, fast or slow, forward or backward).
6. How do you think a mealworm might defend itself against its enemies?
7. What are some ways you used to keep your mealworm on the paper? How did it react?
8. How are mealworms like other animals you know? How are they different?
9. Explain how the hand lenses helped you observe the mealworm.
10. How does the mealworm behave in the bed of peat moss in its holding container compared to when it is on your observation paper or in your observation bag?

Extensions
1. After the students have experienced the various investigations using mealworms, encourage them to look at other animals such as earthworms, crickets, ladybugs, praying mantises, or pill bugs.
2. Make a tube of brown construction paper or use a toilet paper tube or 1/3 of a paper towel tube. Bend six cotton swabs to make jointed legs and glue them to the tube. Decorate to look like a mealworm. Use the "mealworm tube" as a carrier to take home a student's journal or recording sheet about mealworms.
3. Make an album or illustration sheet of all the students' drawings of mealworms.

Curriculum Correlation
Mason, Adrienne. *Mealworms: Raise Them, Watch Them, See Them Change*. Kids Can Press. Buffalo, NY. 2001.

Schaffer, Donna. *Mealworms*. Bridgestone Books. Mankato, MN. 1999.

Resources
If you are not able to find mealworms in a local pet store or bait shop, King mealworms are available from the following sources:

Bassett's Cricket Ranch
1-800-634-2445
http://www.bcrcricket.com

Carolina Biological Supply Company
1-800-334-5551
http://www.carolina.com

Insect Lore
1-800-LIVE BUG
http://www.insectlore.com

Timberline
1-800-423-2248
http://www.timberlinefisheries.com

Home Link
It is not advisable to send mealworms home with students except as pet food (see *Management 7*).

* Reprinted with permission from *Principles and Standards for School Mathematics*, 2000 by the National Council of Teachers of Mathematics. All rights reserved.

My Mealworm Journal

By _____

Mealworms Measure Up!

Pick one of your mealworms to measure. Use this ruler or one like it.

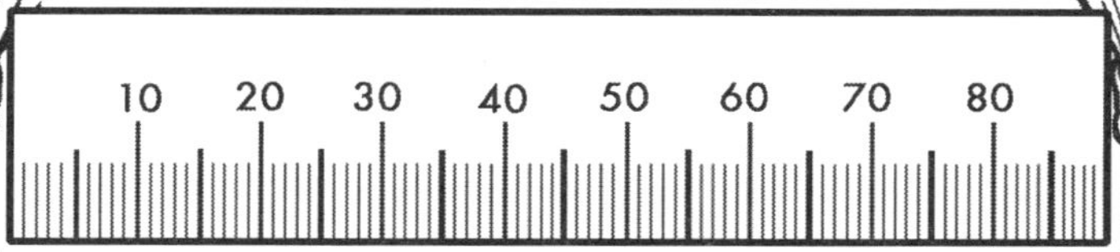

Things longer than my mealworm.

Things shorter than my mealworm.

Mealworm #1

My mealworm has:
- ____ body segments
- ____ legs
- ____ antennae

Mealworm #2

My mealworm has:
- ____ body segments
- ____ legs
- ____ antennae

My mealworms are similar to / different from (circle one) each other.

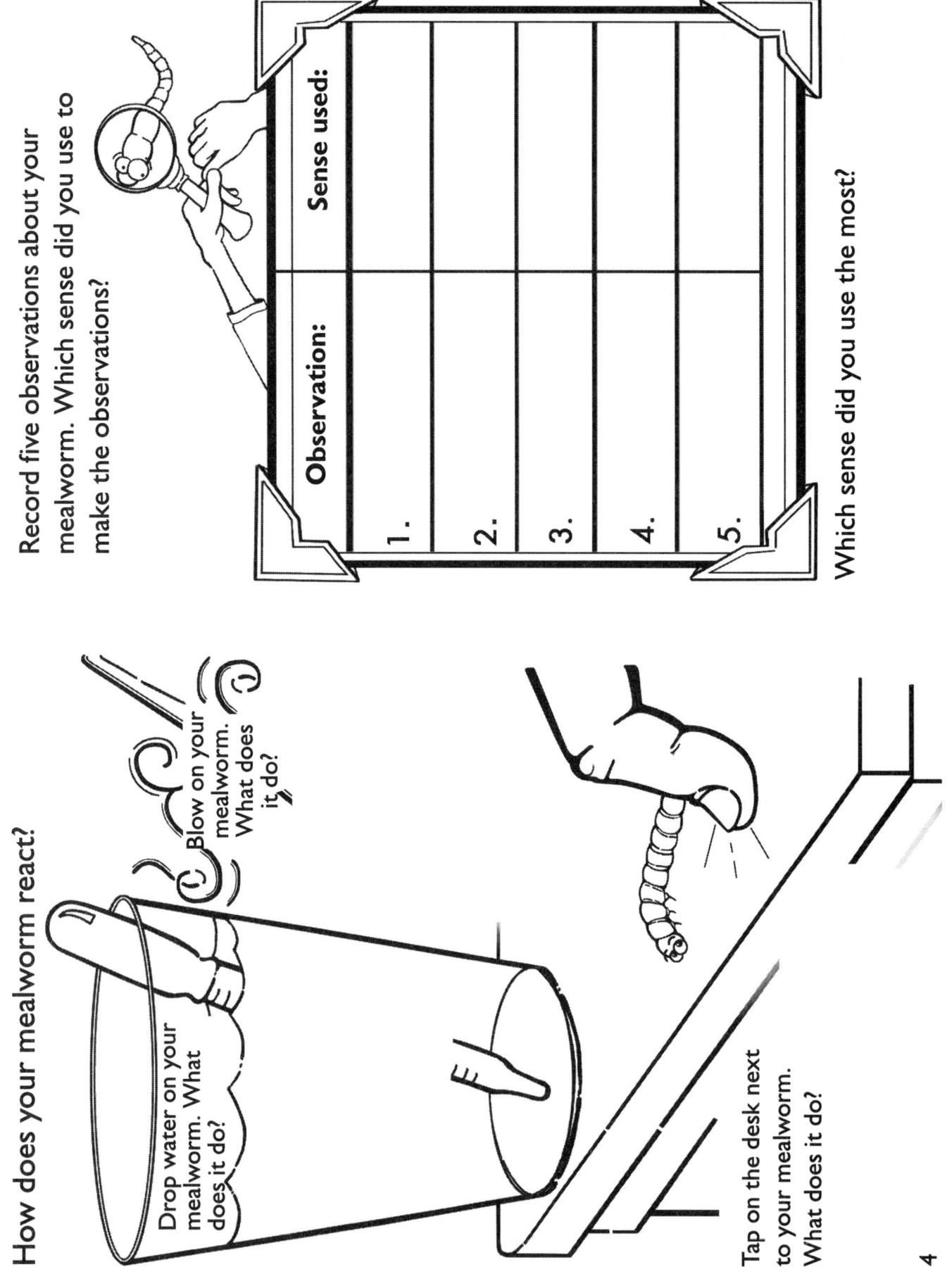

Parts of a Mealworm

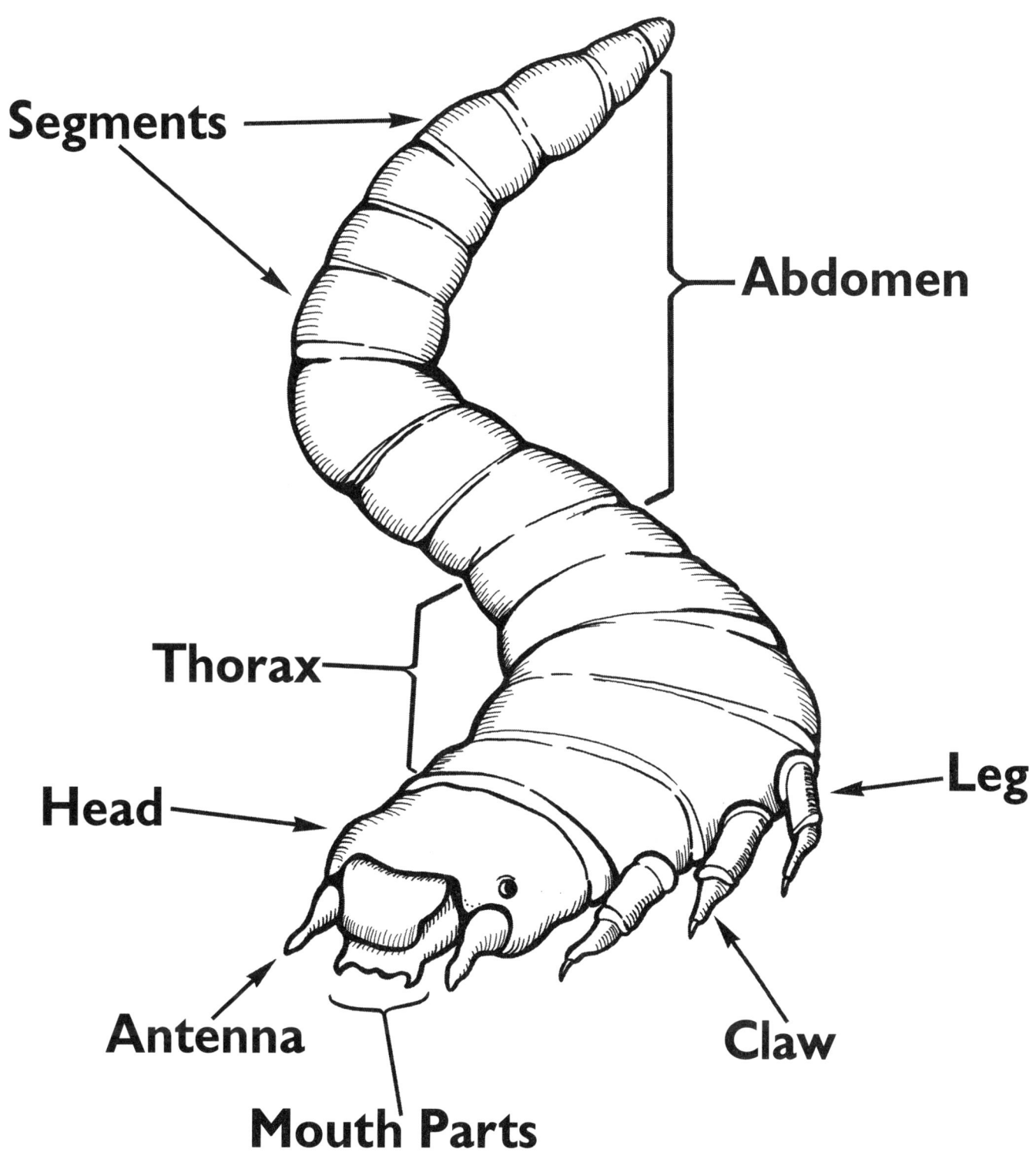

Mealworm Moments, Part Two

Topic
Animal behavior

Key Question
Using observation and the scientific method, what can you learn about mealworms?

Learning Goal
Students will learn about animal behavior using the scientific method to investigate mealworms.

Guiding Documents
Project 2061 Benchmark
- *Things move in many different ways, such as straight, zigzag, round and round, back and forth, and fast and slow.*

NRC Standards
- *Employ simple equipment and tools to gather data and extend the senses.*
- *Communicate investigations and explanations.*
- *The behavior of individual organisms is influenced by internal cues (such as hunger) and by external cues (such as a change in the environment).*
- *An organism's patterns of behavior are related to the nature of that organism's environment, including the kinds and numbers of other organisms present, the availability of food and resources, and the physical characteristics of the environment. When the environment changes, some plants and animals survive and reproduce, and others die or move to new locations.*

NCTM Standards 2000*
- *Use tools to measure*
- *Pose questions and gather data about themselves and their surroundings*

Science
Life science
    animal behavior
        mealworms

Math
Measurement
    linear
    time
    mass
Data collection
Counting

Integrated Processes
Observing
Comparing and contrasting
Collecting and recording data
Communicating
Interpreting data
Generalizing
Predicting

Materials
For the class:
    plastic container (see *Management 1*)
    peat moss (see *Management 1*)
    pieces of apple, zucchini, carrot, or potato
        (see *Management 2*)
    petroleum jelly or transparent tape
        (see *Management 3*)
    timer (see *Management 7*)
    see *Science Investigation Cards* for other
        materials needed

For each group of students:
    Mealworm Moments recording sheets

Background Information
These task card activities continue the study of mealworms begun in *Mealworm Moments, Part One*. The tasks are designed around the scientific method of investigation. After the students have experienced these investigations, encourage them to look at other animals such as earthworms, crickets, ladybugs, praying mantises, or pill bugs that are commonly found on many school playgrounds.

See *Mealworm Moments, Part One* for additional information about mealworms. King mealworms (*Zophobas morio*) are recommended for these activities.

Management
1. When the King mealworms are not being used in an activity, keep them in a plastic container with sides at least 10 cm above a 3-5 cm bed of damp peat moss. They prefer temperatures of 70 to 80 degrees Fahrenheit and should never be kept below 60 degrees.
2. Always keep several pieces of apple, zucchini, potato, or carrot in the container to provide food and moisture. Replace every two to three days to avoid mold.

CRITTERS   48   © 2004 AIMS Education Foundation

3. King mealworms can climb the sides of cardboard containers. To avoid escapes during the investigations, run transparent tape around the inside top edge of the shoeboxes, or put a 2-3 cm wide layer of petroleum jelly just below the edge. Do not leave mealworms unattended in a cardboard box, as they may eat through the cardboard.
4. Mealworms can be purchased from a pet store, bait shop, or other supplier (see *Resources*).
5. Encourage the students to handle the mealworms gently and with respect. Assure those who are reluctant that mealworms will not bite them. Any scratching sensation students feel comes from the mealworms' claws, which are not big enough to damage human skin.
6. It is normal for a healthy mealworm to be inactive at times. If it is dead, it will become stiff and turn dark brown or black.
7. For younger students, supply a one-minute sand timer to use instead of a clock when timing their investigations. A stopwatch is suggested for use by older students.
8. Prior to beginning the lesson, decide how you will use the *Science Investigation Cards*. One way is for each small group of students to use a different task card. Another is to have the students rotate through various stations where they will use a different task card at each station. A third method would be to guide the whole class through each task card at the same time.
9. For *Science Investigation Card 6*, you may wish to construct the mealworm racetrack for students ahead of time. To make the track, fold a 12" x 18" sheet of construction paper into sixths lengthwise. Make three cuts across the width of the paper in the approximate locations indicated in the diagram. (The two cuts next to each other should be far enough apart for a mealworm to fit between them.) The cuts should go to within 1 cm of each edge.

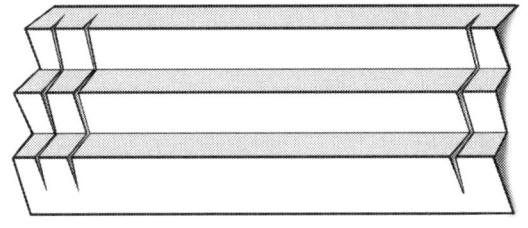

10. One *Mealworm Moments* recording page template has been provided. Decide if you want the students to write in the *Title* and *Question*. If not, write the *Science Investigation Number*, the *Title*, and the *Question* on a separate recording page for each task card before duplicating for each group of students.

11. Be aware that mealworms are a potential pest to agriculture and should never be released into the wild. When you are finished with your mealworms, share them with another class or donate them to your local zoo or pet store.

Procedure
1. Review the charts, the information in the *My Mealworm* recording books, and discussions from *Mealworm Moments, Part One*. Challenge the students to continue their investigation of mealworm behavior using the set of *Science Investigation Cards*.
2. Discuss the *Science Investigation Cards* and how the students will be using them. It is important to stress that students repeat the investigations several times before they come to conclusions.
3. Explain that not all scientific investigations will be conclusive. Predictions may be proven correct or incorrect. Students will learn that many variables contribute to the results of each test. The processes used in these investigations are more important than the results of the actual tests.
4. Direct the students to record the results of each investigation on the appropriate recording pages.
5. Once all investigations have been completed, discuss the *Question* listed on each card and the results.

Connecting Learning
1. What surprised you about the way your mealworm behaved? In what ways did the mealworms behave the way you expected?
2. What else would you like to find out about your mealworm?
3. What other animals might have this same type of behavior? Why do you think they might behave the same as mealworms?
4. What are you wondering now?

Extension
Repeat these investigations using other types of small animals such as earthworms, crickets, or ladybugs.

Curriculum Correlation
Mason, Adrienne. *Mealworms: Raise Them, Watch Them, See Them Change*. Kids Can Press. Buffalo, NY. 2001.

Schaffer, Donna. *Mealworms*. Bridgestone Books. Mankato, MN. 1999.

Resources

If you are not able to find mealworms in a local pet store or bait shop, King mealworms are available from the following sources:

Bassett's Cricket Ranch
1-800-634-2445
http://www.bcrcricket.com

Carolina Biological
1-800-334-5551
http://www.carolina.com

Insect Lore
1-800-LIVE BUG
http://www.insectlore.com

Timberline
1-800-423-2248
http://www.timberlinefisheries.com

Home Link

It is not advisable to send mealworms home with students except as pet food (see *Management 10*).

* Reprinted with permission from *Principles and Standards for School Mathematics*, 2000 by the National Council of Teachers of Mathematics. All rights reserved.

Science Investigation Card 1
Mealworms on the move

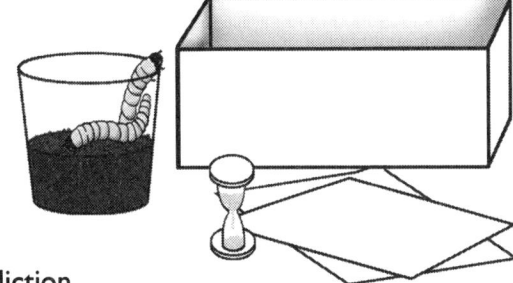

Question
Do our mealworms usually move to a corner, to an edge, climb the walls, or stay in the middle of a box?

Materials
A shoe box, box lid, or similar container
Container of King mealworms
One-minute timer
Recording sheets

Procedure
1. Discuss the question with your group and make a prediction.
2. From the container of mealworms, choose one that seems to be active and place it in the middle of the box. Start the timer.
3. At the end of each minute, record the position of the mealworm in the box: corner, edge, up the wall, or in the center. Leave it there.
4. Start the timer again. After another minute, record the position again. Repeat 10 times.
5. What part(s) of the box does the mealworm seem to prefer to go to or stay in? Why do you think it goes there? Record your findings.

Science Investigation Card 2
Mealworms and Surface Colors

Question
Do our mealworms prefer a white or a black surface?

Materials
A shoe box or similar container
Container of King mealworms
1 piece of black paper, half the size of the bottom of the box
1 piece of white paper, half the size of the bottom of the box
Stopwatch
Recording sheets

Procedure
1. Discuss the question with your group and make a prediction.
2. Tape a piece of black paper and a piece of white paper side-by-side on the floor of the box.
3. Place a mealworm in the middle, start the timer, and watch where it goes. After five minutes, record the location of the mealworm.
4. Put the mealworm back in the middle and repeat *Procedures 2-3* two more times.
5. Based on what you have observed, do you think mealworms like white, black, or have no preference? Record your findings.

Science Investigation Card 3
Mealworms, Wet or Dry?

Question
Do our mealworms prefer a wet or a dry surface?

Materials
Plastic tub or similar container
Paper towel
Container of water
Container of King mealworms
Eyedropper
Recording sheets
Stopwatch or clock

Procedure
1. Discuss the question with your group and make a prediction.
2. Place a sheet of paper towel on the bottom of the tub.
3. Use an eyedropper to carefully wet one end of the paper towel. Add water a little at a time until one end is fairly wet (but not puddled) and the other end is dry.
4. Place four mealworms in the center of the paper towel. Observe the mealworms for 15 minutes and record where the mealworms go and where they end up. Use either a stopwatch or a clock to keep the time.
5. Repeat *Procedure 4* with several different mealworms.
6. What did you find out? Record your findings.

Science Investigation Card 4
Mealworm Hide and Seek

Question
How many mealworms seek shelter under a covered lid compared to a clear lid?

Materials
Plastic tub
Container of King mealworms
Peat moss
Two identical lids—one transparent and one covered with dark material
Clock
Recording sheets

Procedure
1. Discuss the question with your group and make a prediction.
2. Place a 5-cm layer of peat moss in a plastic tub.
3. Place a transparent lid on top of the peat moss on one end of the tub and another lid covered with a dark piece of material (black construction paper) at the other end of the tub.
4. Place 10 mealworms in the peat moss between the two lids.
5. Leave the box for 15-20 minutes using a clock to keep time. When the time is up, lift both lids count and record the number of mealworms under each lid. Count and record the number of mealworms that were not under any lid.
6. What did you find out? Record your findings.

Science Investigation Card 5
Mealworm Munch

Question
How much zucchini do our mealworms eat overnight or over several days?

Material
2 plastic containers
Container of King mealworms
Peat moss
2 pieces of zucchini of equal mass

Procedure
1. Discuss the question with your group and make a prediction.
2. Place a 5-cm layer of peat moss and 10 King mealworms in one container. Place a 5-cm layer of peat moss only in the other container.
3. Use a balance to check that the pieces of zucchini are of equal mass. Ask your teacher to trim the pieces of zucchini, if necessary.
4. Place one piece of zucchini in each container of peat moss. The next day, remove the zucchini pieces and compare their masses.
5. Put the zucchini back in the containers and compare the masses on each of the next two days.
6. What did you find out? Record your findings.

Science Investigation Card 6
Mealworm Marathon

Question
How does your mealworm behave when it is "running a race"?

Materials
12" x 18" piece of construction paper folded and cut
3 index cards
Container of King mealworms
Recording sheets
Stopwatch, optional

Procedure
1. Discuss the question with your group and make a prediction.
2. If your race track is not already made, fold and cut the paper according to your teacher's instructions.
3. Put an index card in each slot. Place a mealworm in each groove of the race track between the two index cards at the start. When both mealworms are lined up evenly, remove the starting gate. The first one to touch the finish line is the winner.
4. If you have a stopwatch, use it to record the race time for each mealworm.
5. Observe the direction the mealworm travels. Which one is faster? Did both finish?
6. Repeat the race using different mealworms or challenge the faster mealworm to race a new mealworm each time.
7. What did you find out? Record your findings.

CRITTERS

Mealworm Moments, Part Two

Science Investigation—Mealworms on the Move

Record your group's results in the table.

Time	Center	Edge	Corner	Up the wall
Start				
1st				
2nd				
3rd				
4th				
5th				
6th				
7th				
8th				
9th				
10th				

Mealworm Moments, Part Two

Science Investigation # _____

Title:

Question:

Our Prediction:

What was observed:

What we think we found out:

What we still wonder about:

Mealworm Moments, Part Two

Science Investigation # _____

Title:

Question:

Our Prediction:

What was observed:

What we think we found out:

What we still wonder about:

Move Along, Mealworm

Topic
Mealworms

Key Question
How does a mealworm walk?

Learning Goals
Students will:
1. observe and describe the way a mealworm walks, and
2. reenact the walking pattern of a mealworm.

Guiding Documents
Project 2061 Benchmark
- A lot can be learned about plants and animals by observing them closely, but care must be taken to know the needs of living things and how to provide for them in the classroom.

NRC Standards
- Each plant or animal has different structures that serve different functions in growth, survival, and reproduction. For example, humans have distinct body structures for walking, holding, seeing, and talking.
- Ask a question about objects, organisms, and events in the environment.
- Employ simple equipment and tools to gather data and extend the senses.

Science
Life science
 animals
  mealworms

Integrated Processes
Observing
Predicting
Collecting and recording data
Comparing and contrasting
Generalizing

Materials
Several mealworms for each group of three students (see *Management 6*)
Containers for mealworms (see *Management 8*)
Mirrors or clear petri dishes
Hand lenses
Crushed bran flakes or oatmeal
Apple slices

Background Information
Mealworms are the larvae of grain beetles (*Tenebrio molitor*), also known as darkling beetles. Grain beetles have the four life stages typical of many insects: egg, larva, pupa, adult. They spend a comparatively long period of their life in the larval (mealworm) stage, and are most often found in mills and feed stores, scavenging on damp, decaying grain and flour. However, they have also been known to congregate in such unlikely places as bags of fertilizer, tobacco, and ground pepper.

Since mealworms are insect larvae, they have six legs arranged in three pairs of two. They also have an *anal leg* that is structurally different and does not figure in this activity, but the students may ask about it. Surprisingly, mealworms usually move each pair of legs as a unit rather than using a left-right motion that most people expect. They also normally move pairs of legs in a 1-3-2 pattern that looks somewhat like a bunny-hop when demonstrated by the students.

Management
1. If the students have not worked with mealworms before, they will need some open-ended time to observe and handle specimens. However, do not emphasize observing the walking patterns before doing this activity.
2. This is a very active lesson. When the entire class is perfecting the mealworm walk, it's best to be outdoors or in a large space. Individual demonstrations are fine in the classroom.
3. Have the students work in groups of three. Each threesome should have its own container of mealworms.
4. Mealworms cannot be counted on to walk on command. It's best to have a backup supply in case they are needed as reinforcements.
5. Because mealworms are cold-blooded, their activity level is directly related to the surrounding temperature. A sluggish mealworm may sometimes be activated by being held gently in a cupped hand for a few minutes.
6. Pet stores stock mealworms as a food source for various reptiles, amphibians, small rodents, fish, and birds. They are also available from many suppliers on the Internet (see *Resources*). Mealworms may be ordered ahead of time and kept in covered containers in the refrigerator until you are ready to use them. They will not grow or be active until they are moved to a warmer temperature.
7. A mealworm colony needs crushed bran flakes or oatmeal and an occasional apple slice for moisture (remove the apple before it gets moldy). The life cycle will continue and the mealworms will thrive for years.
8. Plastic margarine containers with lids make excellent mealworm containers. Keep the mealworms in their containers whenever they are not being directly observed. They do not climb slick sides, and adult meal beetles do not fly.

CRITTERS © 2004 AIMS Education Foundation

9. **Be aware that mealworms are a potential pest to stored grain and feed. Under no circumstances should mealworms or beetles be released into the wild.**

Procedure
1. Ask the students to think about how a mealworm walks. Clarify the number and arrangement of legs. Have each group brainstorm and list all the possible ways that a mealworm might move those three pairs of legs.
2. Put three students in a line with hands on the shoulders of the person in front of them. They are now a six-legged mealworm. Have them demonstrate one of their ideas of how a mealworm might walk. Ask different groups to show other possibilities.
3. You may wish to let all the groups try out all the possibilities they can think of. If so, this is best done outside or in a large space.
4. Have the students discuss in their groups and predict which of the possibilities they think is actually the way a mealworm moves its legs.
5. After the predictions are recorded, have the students observe the mealworms in motion. They will be able to see the pattern of walking by watching the mealworm walk across a mirror or by looking up through a piece of glass (or transparent plastic) as the mealworm walks across it. A petri dish or similar clear container also can be used.
6. When the students have completed their observations and have come to a conclusion, have the groups practice to perfect their mealworm walk. It isn't easy! A natural follow-up is a mealworm-walk race, relay, or an obstacle course.
7. Distribute *Hop Analysis* and have students illustrate the mealworm's walk.

Connecting Learning
1. Does your mealworm always use the same leg pattern when moving?
2. Do all the mealworms walk in the same way?
3. Does the mealworm use the same pattern when it is walking backwards?
4. What other creatures can you think of that might walk in a similar way?
5. What pattern do you think the adult grain beetle uses when it walks? How could you find out?
6. What are you wondering now?

Extensions
1. Have a mealworm race with the real mealworms. Ask the students to come up with creative courses and rules. A simple track can be made by placing two rulers side by side with a narrow space in the middle.
2. Apply this procedure to other critters you study. What would happen if the whole class tried a millipede walk?
3. Use this activity to begin a scientific investigation about the motion of mealworms. Have the students brainstorm what they would like to find out. Possibilities include: Can you make a mealworm walk in a straight line (or turn)? Does a mealworm prefer to go up an incline or down? Why do mealworms sometimes walk backwards?
4. Challenge students to make a flip book of the mealworm's walk.

Curriculum Correlation
Literature
Mason, Adrienne. *Mealworms: Raise Them, Watch Them, See Them Change.* Kids Can Press. Buffalo, NY. 2001.

Schaffer, Donna. *Mealworms.* Bridgestone Books. Mankato, MN. 1999.

Math
1. Graph the results of mealworm races in terms of time and distance.
2. Find out how far different mealworms will travel in a minute. Calculate their speed in centimeters (or inches) per minute. Determine the average speed of mealworms in the class.

Physical Education
Incorporate the Mealworm Walk into your regular PE activities. It makes a wonderful relay event.

Language Arts
Write a related story about a mealworm who dared to walk to a different beat, or a mealworm who had a hard time keeping up, or a mealworm with six left feet, etc.

Music
1. Try doing the *Mealworm Walk* to music. Which music works best?
2. Create a new mealworm dance requiring six legs in motion.
3. Use the *Mealworm Hop* song to reinforce information and practice the *Mealworm Walk.*

Resources
If you are not able to find mealworms in a local pet store or bait shop, King mealworms are available from the following sources:

Bassett's Cricket Ranch
1-800-634-2445
http://www.bcrcricket.com

Carolina Biological
1-800-334-5551
http://www.carolina.com

Insect Lore
1-800-LIVE BUG
http://www.insectlore.com

Timberline
1-800-423-2248
http://www.timberlinefisheries.com

Mealworm Hop

Tune: Bunny Hop
Words: Suzy Gazlay

Music: Ray Anthony and Leonard Auletti

Meal-worms sure walk fun-ny   They're

not like us, it's true   Each

leg pair moves to-geth-er

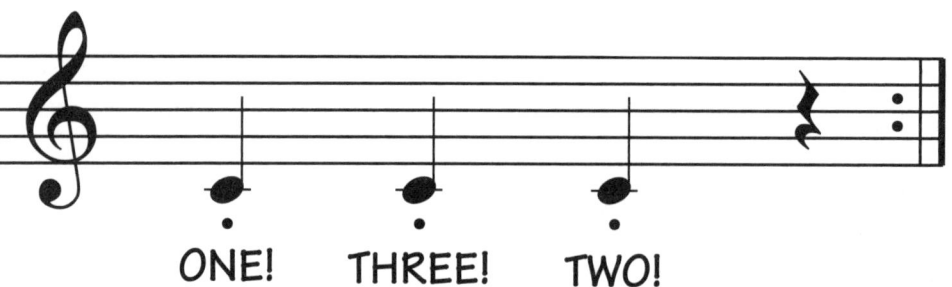

ONE!   THREE!   TWO!

Copyright 1952 Moonlight Music, Inc.
Used with permission

Mealworm Hop

Additional Verses

Their legs don't move in order
But this is what they do:
First, then last, then middle
ONE! THREE! TWO!

Just what are these critters?
Let us give a clue:
They're not worms they're insects
ONE! THREE! TWO!

Mealworms are the larvae
Of beetles, it is true
They live in grain and flour
ONE! THREE! TWO!

Grain beetles have four stages
As lots of insects do:
Egg, larva, pupa, adult
ONE! THREE! TWO!

The eggs are very tiny
And last a week or two
Then hatch out into mealworms
ONE! THREE! TWO!

Mealworms are the larvae
Munching, chomping too
Shedding skins and growing
ONE! THREE! TWO!

The pupae are not active,
Resting, changing too
They don't move like mealworms
ONE! THREE! TWO!

Adults are known as "darkling"
They don't fly, it's true
Do they walk like mealworms?
ONE! THREE! TWO!

Web Threads

Topic
Spider webs

Key Question
Why don't spiders get caught in their own webs?

Learning Goal
Students will perform an investigation that simulates one of the reasons spiders do not get caught in their own webs.

Guiding Documents
Project 2061 Benchmarks
- *Different plants and animals have external features that help them thrive in different kinds of places.*
- *A model of something is different from the real thing but can be used to learn something about the real thing.*
- *One way to describe something is to say how it is like something else.*

NRC Standard
- *Each plant or animal has different structures that serve different functions in growth, survival, and reproduction.*

Science
Life science
   animals
      spiders

Integrated Processes
Observing
Comparing and contrasting
Communicating
Applying

Materials
For each student:
   four 5-cm strips of double-sided transparent tape
     (see *Management 2*)
   four 10-cm strips regular transparent tape
   one cotton swab

Background Information
All spiders produce silk, but not all spiders use their silk to make webs. Only about half of all spiders use silk to trap insects. The silk comes out of the spider's body in liquid form through tiny tubes called spinnerets. The spinnerets are located at the bottom of the spider's abdomen. As the liquid emerges, it hardens, making the silken threads that we see in webs.

Spiders do not usually get caught in their own webs. There are several reasons for this. Spiders can spin different types of silk—sticky and non-sticky, depending on the need. The frame and spokes of most webs are made of non-sticky threads, while the lines going around the web are sticky. The structure of their feet also helps the spiders to travel across the web on the non-sticky threads. If spiders do become stuck on sticky threads, they calmly pull themselves away from the threads rather than thrashing around like many of their victims.

In this activity, students will use double-sided transparent tape to simulate sticky web threads and regular tape (only one side sticky) to simulate the non-sticky threads. They will then investigate what happens when a spider (cotton swab) walks across the different threads.

Management
1. Gather several informational books about spiders to share with the class (see *Curriculum Correlation*).
2. Double-sided transparent tape works best for the sticky threads. If it isn't readily available, roll pieces of tape or attach the strip by bending under a centimeter length at each end.
3. Preparation of the web threads on the student page can be done by the teacher or by the student, depending upon the students' abilities to manipulate the tape. The double-sided tape should be placed on the first web as indicated. The regular transparent tape should be placed sticky side **down** on the radial lines of the second web as indicated.

Procedure
Part One
1. With the class, take a field trip around the school yard in search of spider webs.
2. Caution the students to observe with their eyes only and not to touch any webs they find.
3. Discuss the shape of each web found, what is caught in the web, its location, whether it's still being inhabited by the spider, etc.

CRITTERS   © 2004 AIMS Education Foundation

Part Two
1. Read and discuss some of the literature on spiders and their webs. Focus on information regarding how different spiders use webs to catch their prey.
2. Discuss the fact that when an insect comes in contact with a spider's web, it becomes trapped, stuck on the sticky threads.
3. Invite the students to discuss why they think spiders can walk on their webs and not get stuck like the insects.
4. Tell them that they are going to simulate a spider's web using sticky transparent tape.
5. Direct the students to place the tape on the student pages as described in *Management 3*.
6. Tell students that the cotton swab represents the legs of insects and spiders. Invite them to take one end of the cotton swab and "walk" it along the tape on the first web.
7. Talk about how the cotton sticks to the tape and how it is difficult to move it across the tape. Discuss how this is similar to what can happen to an insect as it becomes stuck to the sticky threads of the spider's web.
8. Describe and discuss how spiders avoid walking on these sticky threads.
9. To simulate this, direct the students to use the other end of the cotton swab to walk the radial lines of tape on the second web and to note any differences.
10. Have students draw an insect on the first web to represent how the insect gets caught in the web.
11. Have them draw a spider on the second web.

Connecting Learning
1. What did you notice about the cotton swab when you walked it along the tape the first time?
2. What did you notice about the cotton swab when you walked it along the tape on the second web?
3. How are the webs alike and how are they different? [Both webs have tape, but only one has sticky tape.]
4. Which threads (tape) are sticky on a spider's web? [Those that go around.] Which ones are not sticky? [Those that go from the center outward.]
5. How is this activity like a spider and an insect in a web? How is it different?
6. Why don't spiders get caught in their own webs?
7. What are you wondering now?

Extension
Students can make spiders and insects from clay, using cotton swabs for their legs and feet.

Curriculum Correlation
Bailey, Jill. *How Spiders Make Their Webs.* Benchmark Books. New York. 1997.

Facklam, Margery. *Spiders and Their Web Sites.* Little, Brown and Company. Boston. 2001.

Levi, Herbert W. *Spiders and Their Kin.* St. Martin's Press. New York. 2003.

Lovett, Sara. *Extremely Weird Spiders.* Avalon Travel Publishing, John Muir Publications. New York. 1996.

Parsons, Alexandra. *Amazing Spiders.* Alfred A. Knopf. New York. 1990.

Ross, Michael Elsohn. *Spiderology.* Carolrhoda Books, Inc. Minneapolis, MN. 2000.

Schnieper, Claudia. *Amazing Spiders.* Carolrhoda Books, Inc. Minneapolis, MN. 2003.

Web Threads

Web 2

CRITTERS

Topic
Fish respiration

Key Questions
1. How does a fish breathe?
2. How many breaths does a fish take in 30 seconds?

Learning Goal
Students will observe the breathing patterns of a goldfish.

Guiding Documents
Project 2061 Benchmarks
- Scientist's explanations about what happens in the world come partly from what they observe, partly from what they think. Sometimes scientists have different explanations for the same set of observations. That usually leads to their making more observations to resolve the differences.
- For any particular environment, some kinds of plants and animals survive well, some survive less well, and some cannot survive at all.

*NCTM Standards 2000**
- Represent data using tables and graphs such as line plots, bar graphs, and line graphs
- Collect data using observations, surveys, and experiments

Math
Measurement
   time
Graphing
Counting
Whole number operations

Science
Life science
   adaptations
      respiration

Integrated Processes
Observing
Predicting
Collecting and recording data
Interpreting data
Comparing and contrasting
Applying
Generalizing

Materials
For each group:
   "feeder" goldfish (see *Management 1*)
   clear plastic two-liter bottles (see *Management 2*)
   goldfish food

For the class:
   stopwatch or watch with second hand

Optional:
   small bottle of aquarium water conditioner
   small fish net
   fresh water plants such as *Elodea*

Background Information
Almost all fish live in water. There are a few specialized fish that can live out of water for a period of time, but this is an exception. As is true of all animals, fish need oxygen to survive. A few fish have lungs, but most fish breathe by means of gills.

The gills are located on the sides of the fish just behind the head. They are covered by a protective flap called the *operculum*. Generally, gills are elaborately branched or subdivided in such a way that their surface areas are maximized. Just beneath the gills' surface is a lavish supply of capillaries that enhance the gas exchange (intake of oxygen, release of carbon dioxide).

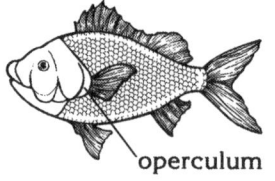

In order to breathe, a fish must take in water, which contains oxygen, through its mouth and over its gills. There the oxygen is absorbed directly into the fish's bloodstream. The stream of water continues on out through the operculum.

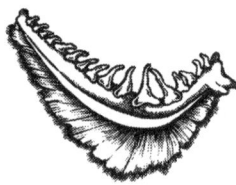

Some fish can enhance the flow of water streaming over their gills by swimming with their mouths open. A few fish are such fast swimmers that their forward movement is sufficient to keep water moving over the gills. Most, like goldfish, take in oxygenated water

CRITTERS   © 2004 AIMS Education Foundation

with a definite, easily observable gulping motion. By counting the number of gulps in a given time period, students can determine the goldfish's respiration rate.

Management
1. Small "feeder" goldfish are available at pet stores and are very inexpensive.
2. Use a clear plastic two-liter bottle with the label removed for each group of students. Use a utility knife or scissors to cut off the neck of the bottle. Fill the bottle three-fourths full of water and add a small goldfish. Put in a piece of water plant, if available.
3. Students should be encouraged to record their observations throughout this activity using both words and drawings.
4. Some things to know about the care and maintenance of goldfish:
   a. Most tap water has chlorine and possibly other chemicals that may harm the fish. Use bottled spring water, or add a few drops of aquarium water conditioner to any new tap water you use. If this is not possible, let the water sit in an open container for at least a full day. Do NOT use distilled water; it has had the oxygen removed and your fish can suffocate.
   b. Goldfish tolerate a wide range of conditions; however, because they are cold-blooded, they may go into shock if they are moved into a container of water that is significantly warmer or colder than where they have been. Always allow all containers of water to arrive at room temperature before changing fish from one container to another. Don't leave fish in a window or too near a source of heat or cold.
   c. When transferring fish, use a net, a small cup, or your hands. Make sure that your hands are thoroughly wet and free of soap before you touch the fish.
5. When timing the gulping rate (respiration), start and stop the students together so they don't have to take their eyes off the fish. If the gulping is especially rapid, use a shorter time period for ease of counting.

Procedure
1. Have students get into groups and give each group a container with a goldfish.
2. Give students time to observe the fish as it gulps the water. Ask them why they think it does this. Lead the discussion to the topic of respiration and gills.
3. Direct the students to count and record the number of gulps their fish make in 30 seconds. Have them repeat this two more times and find the average of the three trials.
4. Ask students to predict what they think will happen to the respiration rate if they excite the fish by moving the water. Record the predictions on the overhead projector or on the chalkboard. Invite students to quantify their predictions.
5. Direct them to gently move the container and immediately count and record the number of gulps in 30 seconds.
6. Have students record both sets of data on the graph. If there are not enough spaces on the graph, invite students to problem solve the solution. Allow them time to compare and contrast the results from the first count and the second count. Discuss.

Connecting Learning
1. What was the average respiration rate of your group's fish for 30 seconds? How did this rate compare to those of other groups? How can you account for the similarities and/or differences?
2. Did moving the water make a difference in the number of gulps per 30 seconds? Is this what you expected? Why or why not? How do you account for what you observed? What similar events might occur in the wild?
3. How does your fish's respiration rate compare to yours? How can you find out?
4. Why does the fish gulp?
5. Why can't you breathe in water like your fish?
6. What are you wondering now?

Extensions
1. Discuss whether a fish's gulping rate would be different when it is asleep or resting. Brainstorm ways to investigate this problem. Have students design and carry out their own strategies for research. Have them compare their results with others and generalize their conclusions.
2. Use a pipette to place a drop of vegetable food coloring directly in front of the mouth of the fish. You should be able to see the dye as it is taken in at the mouth, passed over the gills, and expelled.

Curriculum Correlation
Literature
Parker, Steve. *Eyewitness Books: Fish*. DK Publishing. New York. 2000.

Math
Based upon your observation of the average respiration rate of your fish, calculate its respiration rate for a minute, hour, day, week, year.

Home Link
After a discussion of the proper care of fish, allow the students to take the fish home. Encourage them to explain to other members of their families how a fish gets oxygen from the water.

* Reprinted with permission from *Principles and Standards for School Mathematics*, 2000 by the National Council of Teachers of Mathematics. All rights reserved.

Goldfish Gulps

The respiration rate of your goldfish is the number of breaths it takes in a certain amount of time. (A goldfish is breathing every time it takes a gulp of water.) In this activity, you will measure how many breaths (gulps) your goldfish takes in 30 seconds.

Observe your goldfish and record how many gulps it takes in 30 seconds. Record this information in the first table. Repeat two more times and find the average of the three trials.

Agitate the water by gently shaking the container. Immediately count and record the number of gulps in 30 seconds. Repeat two more times and find the average of the three trials.

Graph the average respiration rate for your fish under normal conditions and when it is agitated.

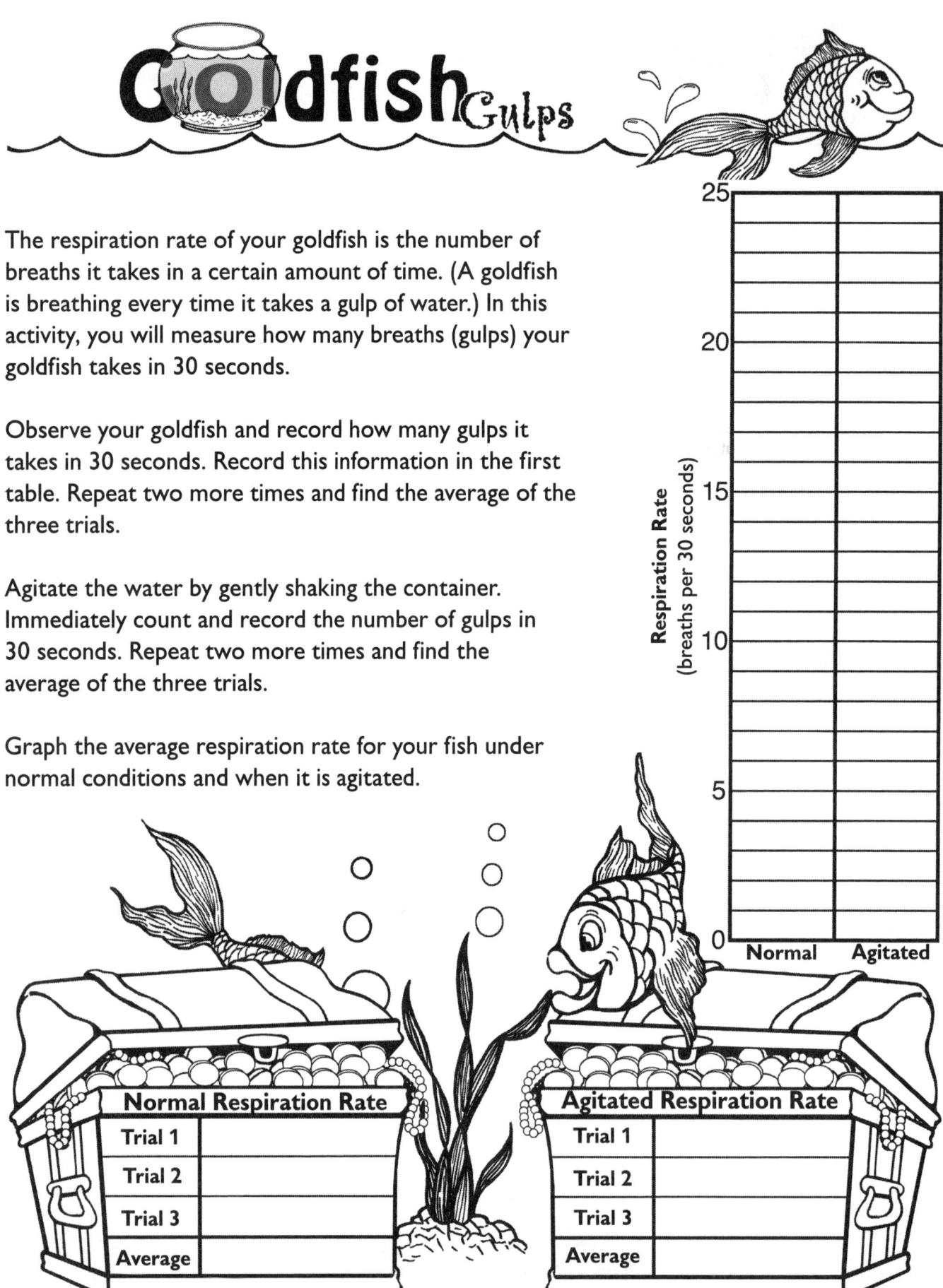

Normal Respiration Rate	
Trial 1	
Trial 2	
Trial 3	
Average	

Agitated Respiration Rate	
Trial 1	
Trial 2	
Trial 3	
Average	

Hot Foot Cold Feet

Topic
Temperature preference of pill bugs

Key Question
What temperature will the pill bugs prefer: hot, room temperature, or cold?

Learning Goals
Students will:
1. guess the number of pill bugs that will be in each temperature section after 10 minutes,
2. set up an experiment to test their guesses,
3. collect and record their data, and
4. draw conclusions about the temperature preferences of pill bugs.

Guiding Documents
NRC Standards
- *The behavior of individual organisms is influenced by internal cues (such as hunger) and by external cues (such as a change in the environment). Humans and other organisms have senses that help them detect internal and external cues.*
- *Plan and conduct a simple investigation.*
- *Use data to construct a reasonable explanation.*
- *Communicate investigations and explanations.*

NCTM Standards 2000*
- *Collect data using observations, surveys, and experiments*
- *Represent data using tables and graphs such as line plots, bar graphs, and line graphs*

Math
Counting
Average
Graphing
Measurement
    time
    temperature

Science
Life science
    animal behaviors
        temperature preferences

Integrated Processes
Observing
Predicting
Collecting and recording data
Controlling variables
Generalizing

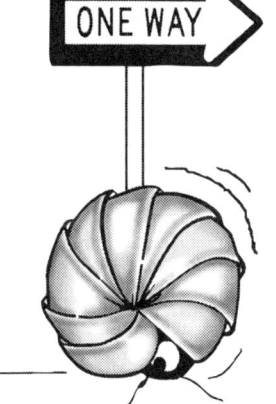

Materials
For each group:
    3 sandwich-size zipper-type plastic bags
    heavy-duty aluminum foil (12" x 36")
    10 pill bugs
    hot water
    ice water
    room-temperature water
    thermometer
    dark colored paper

Background Information
Pill bugs are common garden critters that belong to the order *Isopoda*. They are also known as sow bugs, roly-polies, and wood lice. An isopod is any crustacean that has a segmented body, each segment of which has one pair of similar legs.

Pill bugs are invertebrates. They do not have the ability to control the temperature of their bodies. As a result, their existence is dependent upon the temperature of their surroundings. If the environment becomes too cool or too warm, the pill bug cannot adjust to it and will die. Most invertebrates prefer a temperature range between 60° to 90° F. Pill bugs generally prefer dark, moist, and cool areas (65° F).

Management
1. Collect enough pill bugs beforehand so that each group has 10. The best times to find pill bugs are in the mornings and evenings when it is cooler and damp. Look in flowerbeds and gardens, or other places where there are decomposing plant materials.
2. Heavy duty aluminum foil works best for making the ramps. If you use lightweight foil, you may wish to double it up to increase the sturdiness. Each group will need a sheet about three feet long. To make the aluminum foil ramps, students will need to fold the foil into the shape of a rain gutter. Be sure that the sides are vertical. They should fold up the ends so that the pill bugs will not be able to escape.

3. Be sure to use sturdy plastic bags that have a good seal and will not leak or be easily punctured. Fill the room-temperature bags with water ahead of time to allow them to reach room temperature, but fill the hot and cold bags just before beginning the activity.

CRITTERS   © 2004 AIMS Education Foundation

Procedure
1. Have students discuss what they already know about pill bugs. Record their responses on the board.
2. Ask them to share how they have learned about pill bugs. Many may respond by saying they have observed much of what they know about the animals. Stress the importance of observation.
3. Ask the *Key Question* and state the *Learning Goals*.
4. Explain the setup for the experiment. Demonstrate how to make an aluminum foil ramp, and assist groups as they make theirs. Be sure that everyone has vertical sides and has properly sealed up each end to prevent escapes.
5. Distribute three plastic bags—one filled with ice water, one with room temperature water, and the third with hot tap water—to each group. Remove as much air as possible from each bag.
6. Give groups thermometers and have students find the temperature of each of the bags and record it on the activity sheet in the bags marked *hot, room temperature,* and *cold*.
7. Have groups position their ramps over the bags so that the room temperature bag is in the middle, and the hot and cold bags are on either end. There should be some space between each bag.

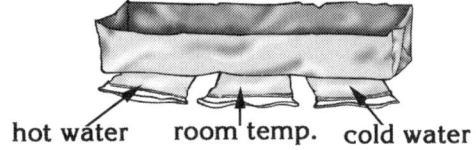
hot water   room temp.   cold water

8. Give each group their 10 pill bugs, and instruct them to carefully place the bugs in the center of the ramp, over the room-temperature bag of water.
9. Explain how to time and count the number of pill bugs in each temperature zone and record it on the activity sheets. Use the bags as a reference point for each temperature zone.
10. Be sure to have the students guess the number of pill bugs that will be in each area at the end of 10 minutes.
11. Have groups begin collecting and recording their data. Be sure that one person in the group is keeping track of the time so that every two minutes a count is made
12. After the 10 minutes are over, have each group cover their ramp with dark colored paper.
13. Let the ramps sit covered for an additional five minutes, then remove the papers and make a final observation and location count.
14. Ask each group to share their final count per temperature zone on their group sheets and graph the data.
15. If appropriate for your students, have them complete the third student page by calculating and graphing the percent of pill bugs in each temperature zone.

Connecting Learning
1. Where were the most pill bugs after 10 minutes? [most of the pill bugs will likely be in the cold section]
2. What does this tell you about the temperature preferences of pill bugs? [they prefer cooler temperatures]
3. Did it make a difference when the ramp was covered versus uncovered? [When it is uncovered, the pill bugs will likely remain active and exploring because of the light. When the ramp is covered, many of the pill bugs may settle down and even curl up, usually in the cold section.]
4. What does this tell you about pill bugs and their response to light?
5. Were the results consistent throughout the room?
6. If not, what are some possible variables that could have caused the differences?
7. What did the tables and graph tell you about the average pill bug? Was this information similar to what your group discovered?
8. What are you wondering now?

Extensions
1. Have students use the pattern included to make their own pill bugs. Copy the page onto gray or brown paper.
2. Repeat the experiment using other "critters" such as earthworms, snails, or mealworms.
3. Have students place equal number of critters in each temperature area and see how many are in that area after five minutes.

Curriculum Correlation
Literature
Ross, Michael Elsohn. *Rolypolyology*. Carolrhoda Books, Inc. Minneapolis, MN. 1996.

Schaffer, Donna. *Pillbugs*. Bridgestone Books. Mankato, MN. 1999.

Language
Write a short commentary on what a small critter might say while choosing the best temperature to live in.

* Reprinted with permission from *Principles and Standards for School Mathematics*, 2000 by the National Council of Teachers of Mathematics. All rights reserved.

Hot Foot Cold Feet

What temperature do pill bugs prefer: hot, room temperature, or cold?

Measure the temperature of the water in each of your baggies. Record this information in the spaces below. Be sure to label your measurements as either Celsius (C) or Fahrenheit (F).

Where do you think most of the pill bugs will be after 10 minutes—in the hot section, the room temperature section, or the cold section? Record your guesses here.

Time	Hot	Room Temperature	Cold
2 minutes			
4 minutes			
6 minutes			
8 minutes			
10 minutes			
+5 minutes (covered)			

Were your guesses correct? Why or why not?

What reasons can you think of for your results?

Hot Foot Cold Feet

Use the table below to record the data from the rest of the class.

Number of critters after 10 minutes

Group	Hot	Room Temperature	Cold
1			
2			
3			
4			
5			
6			
7			
8			
Total			
# of groups			
Average			

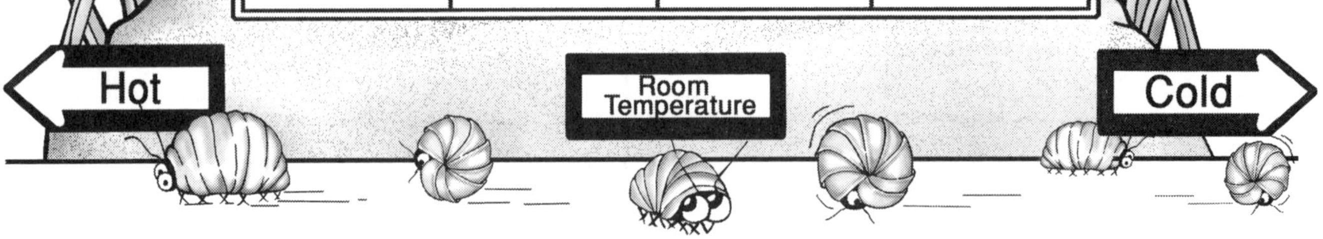

⬅ Hot   Room Temperature   Cold ➡

According to your average results, where were most of the pill bugs located after 10 minutes?

Were results consistent from group to group? What might be some reasons for any differences?

Hot Foot *Cold Feet*

In the table below, determine the decimal value and percent of pill bugs in each of the temperature sections. Make a bar graph of the percentages.

	Average # of critters in each temp.	÷	Average # of critters per group	=	Decimal value	x	100	=	Percent
Hot		÷		=		x	100	=	
Room Temp.		÷		=		x	100	=	
Cold		÷		=		x	100	=	

Percentage of critters in each temperature

100			
95			
90			
85			
80			
75			
70			
65			
60			
55			
50			
45			
40			
35			
30			
25			
20			
15			
10			
5			
	Hot	**Room Temperature**	**Cold**

What does the graph tell you?

 © 2004 AIMS Education Foundation

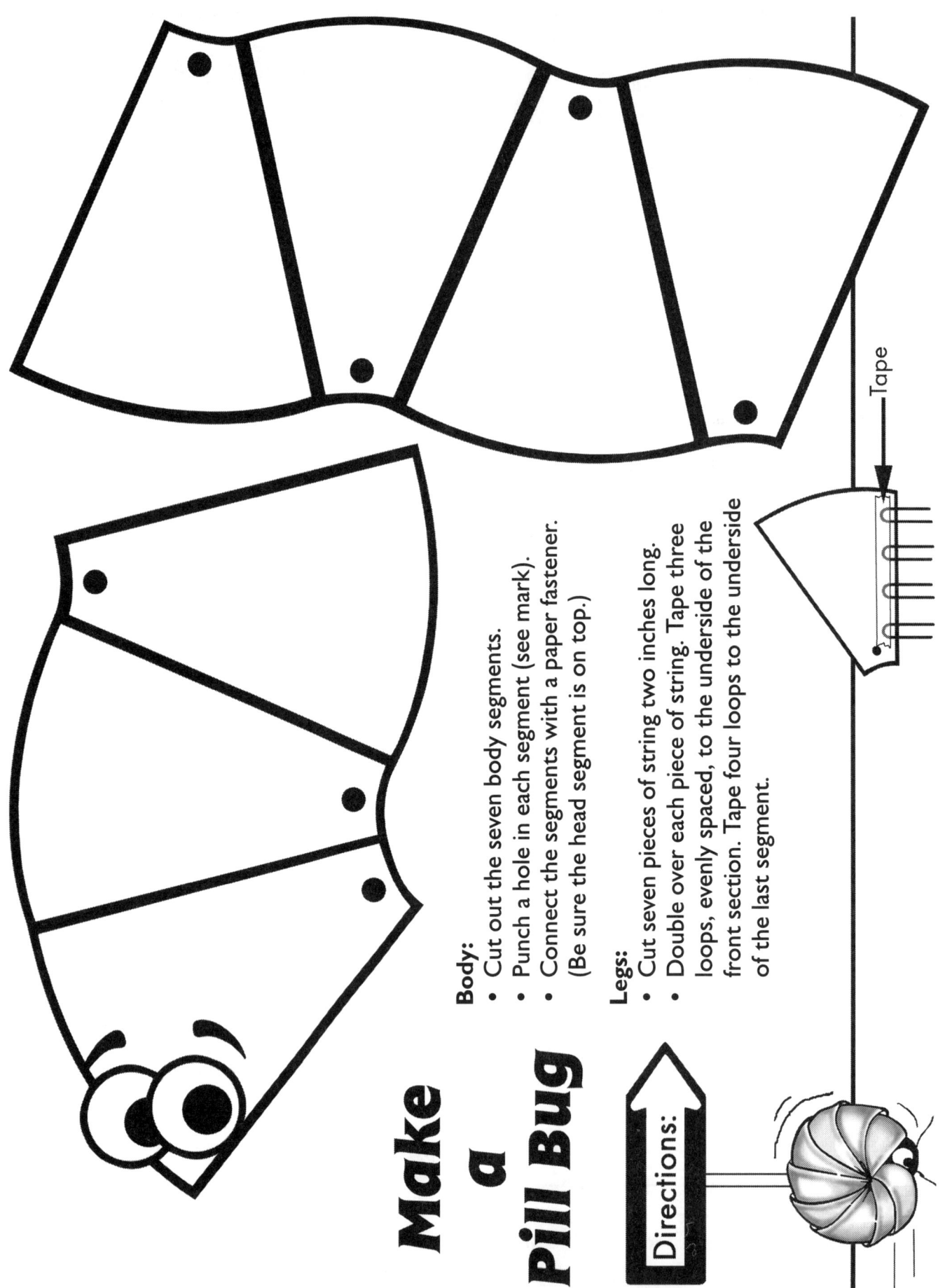

Make a Pill Bug

Directions:

Body:
- Cut out the seven body segments.
- Punch a hole in each segment (see mark).
- Connect the segments with a paper fastener. (Be sure the head segment is on top.)

Legs:
- Cut seven pieces of string two inches long.
- Double over each piece of string. Tape three loops, evenly spaced, to the underside of the front section. Tape four loops to the underside of the last segment.

Tape

CRITTERS © 2004 AIMS Education Foundation

Wings 'n' Webs

Topic
Insects and spiders

Key Question
What are the external differences between the bodies of insects and spiders?

Learning Goals
Students will:
1. compare and contrast the external features of insects and spiders,
2. construct either an insect or a spider using simple materials,
3. identify classmates' critters as either insects or spiders, and
4. make a graph showing the number of insects and spiders constructed.

Guiding Documents
Project 2061 Benchmarks
- A great variety of kinds of living things can be sorted into groups in many ways using various features to decide which things belong to which group.
- A model of something is different from the real thing but can be used to learn something about the real thing.
- One way to describe something is to say how it is like something else.

NRC Standard
- Each plant or animal has different structures that serve different functions in growth, survival, and reproduction. For example, humans have distinct body structures for walking, holding, seeing, and talking.

Math
Counting
Graphing

Science
Life science
    animal characteristics
        insects
        spiders

Integrated Processes
Observing
Comparing and contrasting
Collecting and recording data
Organizing data
Drawing conclusions
Generalizing

Materials
An insect and spider for each group
    (see *Management 2*)
Hand lenses
Materials for making insects and spiders
    (see *Management 3*)
Scissors
Activity sheets
The Critter Connection: Insects
The Critter Connection: Spiders
#19 rubber bands, 2 per student

Background Information
See the spider and insect rubber band books.

Management
1. Have students work together in groups of four to five.
2. Each group of students needs an insect and a spider. It is strongly recommended that real insects and spiders be used. Specimens can be living or dead, but living ones should be in appropriate containers so that they will not escape into the classroom. If real specimens are not possible, use the pictures provided in the investigation.
3. Students will need a variety of materials for constructing their insects and spiders. Make available such items as chenille stems, toothpicks, clay, construction paper, aluminum foil, etc. Encourage students to develop a plan for their critters and to only take the materials they will need.

Procedure
1. Divide students into groups of four or five.
2. Give each student two rubber bands and a copy of the *Critter Connections* pages. Assist them in assembling their rubber band books. Read and discuss the information as a class.

3. Give each group a spider and an insect. If real specimens are not available, give each group a picture of a spider and an insect.
4. Distribute the activity sheets to each student.
5. Have each student record the name of the insect and spider used in his or her group (cricket, daddy long-legs, grasshopper, jumping spider, etc.).
6. Instruct students to compare the insects and spiders and to record their observations in the table on the first student page.
7. Discuss the findings as a class.
8. Ask all students to write a sentence or two about the differences observed.
9. Once students know some of the differences, have them finish the drawings at the bottom of the first page. (These drawings are symmetrical, and can be completed using a reflect/view or by folding the paper in half.)
10. Make available the necessary materials for students to create their insects and spiders and allow time for construction. Each student should make his or her own spider or insect.
11. Have students make drawings of their creations, label them as spiders or insects, and name them.
12. Within their groups, have students guess each other's creations and record their reasoning on the second student page.
13. As a class, build a real graph on the floor of spiders and insects using the students' creations. Have students use the real graph to help them make a bar graph of their own on the final student page.
14. Close with a time of class discussion and sharing.

Connecting Learning
1. What are two ways that insects and spiders are the same? [both invertebrates, both have exoskeleton, both arthropods, etc.]
2. What are four ways that insects and spiders are different? [insects have six legs, spiders have eight; insects have three body parts, spiders have two; most insects have two compound eyes, most spiders have eight simple eyes; most insects go through metamorphosis, spiders do not, etc.]
3. How were you able to tell if your classmates' critters were insects or spiders?
4. What did the graph show us about our creations? [the number of insects and spiders that were made]
5. Have you seen creatures that are not insects or spiders but appear very similar? What do you think they are?
6. What are you wondering now?

Extensions
1. Collect insects and spiders and identify them.
2. Make a booklet of one spider or insect.

Curriculum Correlation
Literature
Allen, Judy. *Are You a Spider?* Kingfisher. New York. 2000.

Gibbons, Gail. *Spiders*. Holiday House. New York. 1994.

Mound, Laurence. *Eyewitness Books: Insect*. DK Publishing. New York. 2000.

Ross, Michael Elsohn. *Spiderology*. Carolrhoda Books, Inc. Minneapolis, MN. 2000.

Schnieper, Claudia. *Amazing Spiders*. Carolrhoda Books, Inc. Minneapolis, MN. 2003.

Creative Writing
Write a make-believe story about a spider or insect.

Geography
What is the territory of certain insects or spiders?

History
Research the following questions:
- How have spiders and insects been used to represent good and evil in past civilizations?
- How have insects caused crop damage, and what was historical significance of this damage?
- Why do states inspect fruits and vegetables at borders?

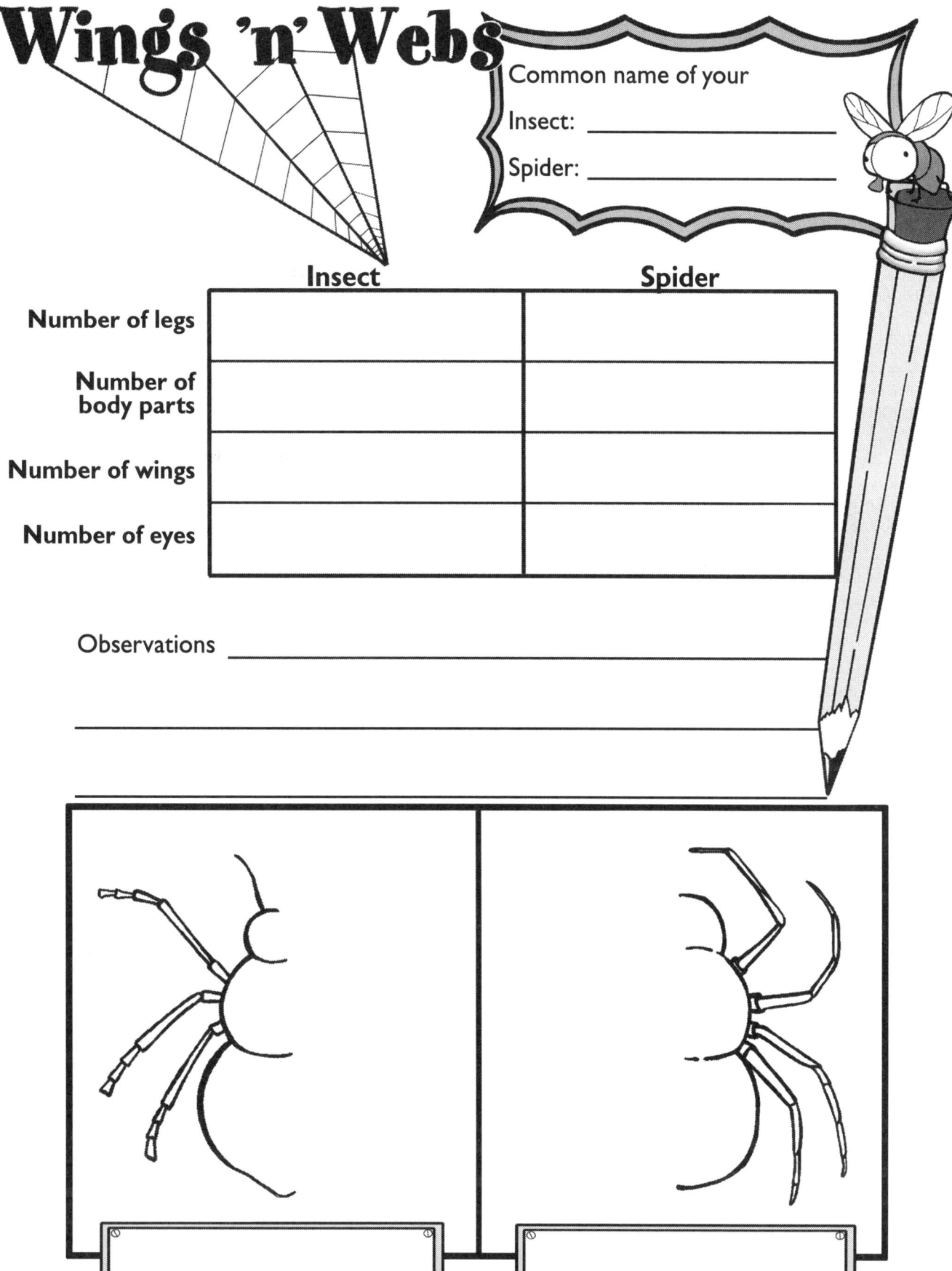

Wings 'n' Webs

Make an insect or spider out of clay, toothpicks, pipe cleaners and construction paper. Don't tell anyone what it is.

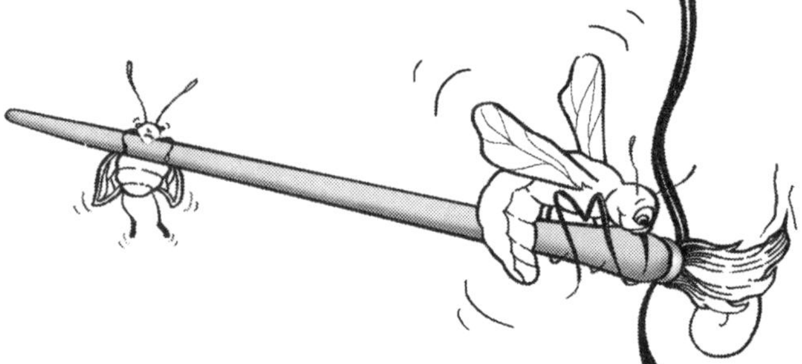

Draw it:

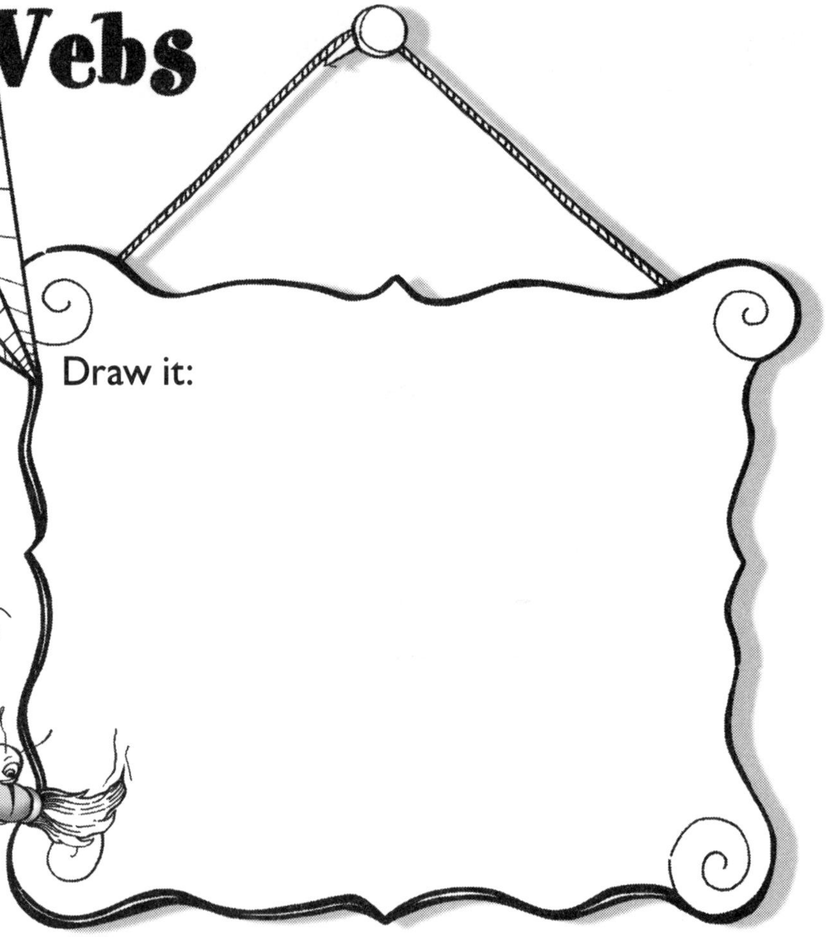

What is it? _____

Give it a name

Observe what each person in your group made:

I think _____'s critter is a _____ because
_____

I think _____'s critter is a _____ because
_____

I think _____'s critter is a _____ because
_____

How many did you guess correctly?

CRITTERS © 2004 AIMS Education Foundation

Wings 'n' Webs

Make a whole-class floor graph of the insects and spiders that you made. Now make your own bar graph from the floor graph.

Animal Antics

Topic
Animal classification

Key Question
How do scientists classify animals?

Learning Goal
Students will sort animals into appropriate classifications in the animal kingdom.

Guiding Documents
Project 2061 Benchmarks
- *A great variety of kinds of living things can be sorted into groups in many ways using various features to decide which things belong to which group.*
- *Features used for grouping depend on the purpose of the grouping.*

NRC Standard
- *Scientists use different kinds of investigations depending on the questions they are trying to answer. Types of investigations include describing objects, events, and organisms; classifying them; and doing a fair test (experimenting).*

NCTM Standards 2000*
- *Represent data using tables and graphs such as line plots, bar graphs, and line graphs*
- *Sort and classify objects according to their attributes and organize data about the objects*

Math
Graphing
Computation
Sorting

Science
Life science
　animal classification

Integrated Processes
Observing
Classifying
Predicting

Materials
Activity sheets
Animal bags (see *Management 2*)
Classification chart (see *Management 3*)
#19 rubber bands (see *Management 4*)

Background Information
　The animal kingdom can be classified into two groups—the vertebrates and the invertebrates. Vertebrates are animals with backbones, and can be classified into five sub-groups: mammals, birds, fish, reptiles, and amphibians. The invertebrates are classified into many groups, but for this activity will only be sorted into four sub-groups: ringed worms, arthropods (insects, spiders, crabs), mollusks (slugs, squid, snails), and echinoderms (spiny-skinned animals like sea stars, sea urchins, and sand dollars). This activity is dealing with classification on an elementary level and is not intended to be complete, but rather to expose elementary students to the idea of classifying animals into groups according to attributes they have in common. Likewise, this activity may be too complex for primary students, but a creative teacher can use the pictures and have students sorting and classifying on a more primary level.

Management
1. This activity can be done as a game as described in the *Procedure*, or simply as a sorting exercise using the animal pictures and the rubber band book to classify animals.
2. Each group of students will need one bag of animals. The bag should contain several of the pictures from the animal picture sheet included with the activity. Bags can be supplemented with some or all of the following: animal cookies, goldfish crackers, gummy worms, or plastic bugs, spiders, snakes, and lizards. No two bags should contain the same combination of animals.
3. Each group will need one copy of the two-page classification chart. It should be cut along the dashed line and taped together to form one large page.

CRITTERS　　　　　81　　　　　© 2004 AIMS Education Foundation

4. Each student will make a rubber band book with information about the animal kingdom. The large rubber bands (size #19) will be used to hold the rubber band books together.

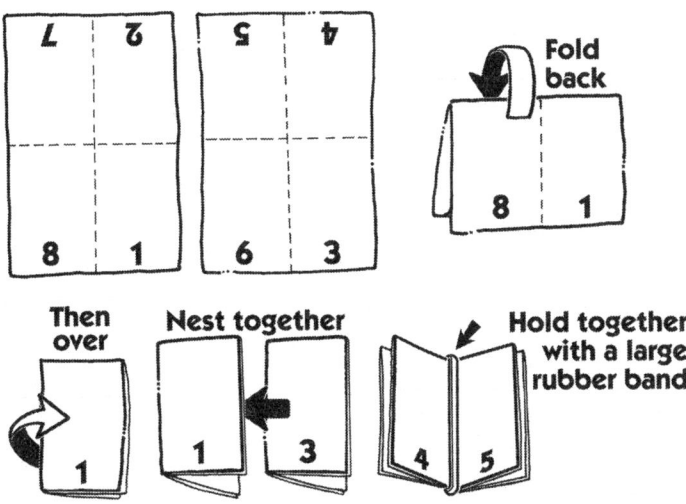

Procedure
1. Give each student a copy of the rubber band book pages and a #19 rubber band. Assist the students in assembling their rubber band books and read through the information as a class.
2. Divide the class into teams of three to four students. Give each team an animal bag. Each team's bag should contain about the same number of animals.
3. Allow the students to look at the animals in their bags for a few minutes and then have them return the animals to the bags.
4. Hand out a classification chart and the activity pages to each group. Explain that the object of the game is to correctly sort the animals according to the classification system given on the chart.
5. Discuss the point system for the game as given on the data sheet. Have groups predict how many points they think they will get and record their predictions on their data sheets.
6. Instruct teams to empty their bags and place all of the animals at the bottom of the classification chart in the space labeled *All Animals*. Have them count the number of animals and record this value on the data sheet.
7. Have groups begin to classify their animals, starting with the division into vertebrates and invertebrates. Check each team for correct placement. One point is received for each correct placement, and should be recorded on the data sheet.
8. Instruct teams to continue sorting their animals into the categories. When they are finished, check each for correct placement and help students record and calculate their point totals. For example, if a team correctly classifies three reptiles, they will receive 3 x 4 (12) points and record those points on the data sheet.
9. After each team has calculated its score, record and graph the scores for all of the teams on the graphing page.

Connecting Learning
1. What do the mammals have in common? ...the reptiles? ...the mollusks, etc.?
2. Was your team's predicted score close to your actual score? Why or why not?
3. Which group of animals is most often selected as pets? Why do you think this is so?
4. How would you sort the mammals into smaller categories?
5. Can you devise an animal classification system that is different than the one given on the chart? Explain.
6. What animals would you like to add to the chart? Where would they fit?
7. Can you think of another animal that would fit into each group? Name those animals and the groups into which they fit.
8. What are you wondering now?

Extensions
1. Draw a large classification chart on butcher paper and place it on a bulletin board. Have each student cut out and color an animal picture and paste it on the chart in the appropriate place.
2. Have older students find the ratios or percents of animals they had in each category on the chart.
3. Have younger students do the activity with only the animal pictures and sort them using a variety of methods of their own devising.

Curriculum Correlation
Art
Use the cut-out animals to make a zoo collage.

Creative writing
Have students write an animal story. Rudyard Kipling's *Just So Stories* are great for ideas. You can read "The Elephant's Child" story from this book about how the elephant got its trunk and then have students write their own stories about how other animals got their unique features.

Research
Students can research other animal classification systems or find other invertebrate classes that were not included in the chart in this lesson.

* Reprinted with permission from *Principles and Standards for School Mathematics*, 2000 by the National Council of Teachers of Mathematics. All rights reserved.

Vertebrates and Invertebrates

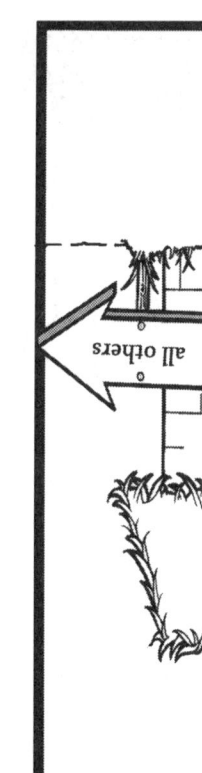

Animals can be classified into two groups. The *vertebrates* are animals with backbones. The *invertebrates* are

Human Boy (vertebrate)

animals without backbones. Run your hand down your back. Do you feel the bumpy bones? That is your backbone.

You are a vertebrate. Only about 5% of all the animals on Earth are vertebrates.

There are many groups of invertebrates. Here are four of the main ones:

- **Annelids:** cold-blooded animals that have soft bodies with sections
- **Echinoderms:** cold-blooded animals that have bodies with rough skin and sharp spines
- **Mollusks:** cold-blooded animals with a soft body and sometimes a hard shell
- **Arthropods:** cold-blooded animals with jointed legs

Animals

Vertebrates

mammals
humans
dogs
rabbits
lions
cats
whales

amphibians
frogs
toads
salamanders

reptiles
turtles
snakes
lizards
alligators

fish
trout
bass
eels

birds
eagles
turkeys
sparrows

Invertebrates

mollusks
clams
oysters
snails
octopuses
squid

arthropods
spiders
insects
crabs
lobsters

echinoderms
sea star
sea urchins
sand dollars

annelids
earthworms
clamworms
leeches

Invertebrates are far more numerous than vertebrates. Of all the animals that have been discovered on Earth, about 95% are invertebrates. The largest group of invertebrates is the arthropods, which includes insects. Insects make up about 75% of all known animal species.

There are five groups of vertebrates:
- **Mammals:** warm-blooded animals that have hair or fur and are born alive
- **Birds:** warm-blooded animals that have feathers and lay eggs
- **Fish:** cold-blooded animals that have scales, gills, and fins and lay eggs
- **Reptiles:** cold-blooded animals that have scales and lungs and lay eggs
- **Amphibians:** cold-blooded animals that have smooth skin and can live on land or in water

Warm-blooded animals have constant body temperatures. Cold-blooded animals have body temperatures that adjust to the temperatures of their environments.

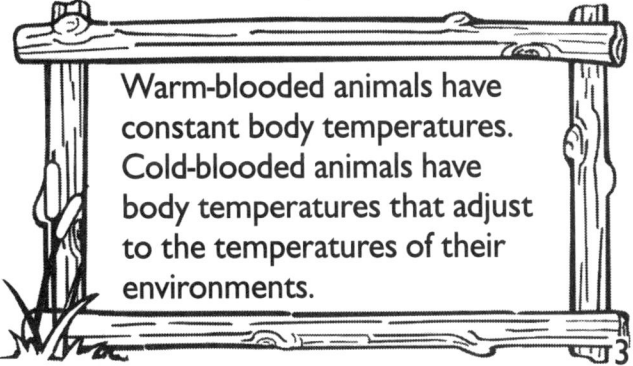

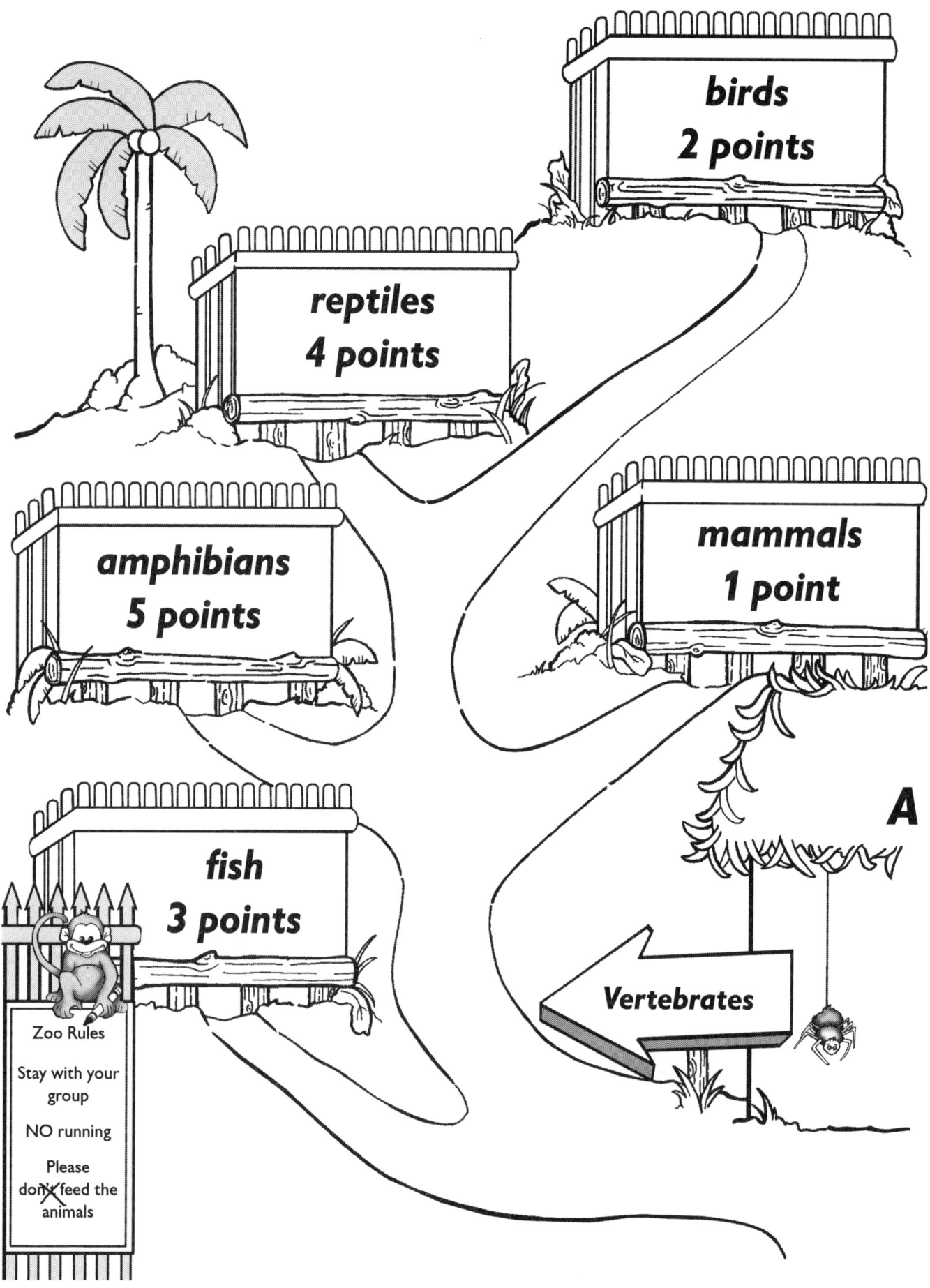

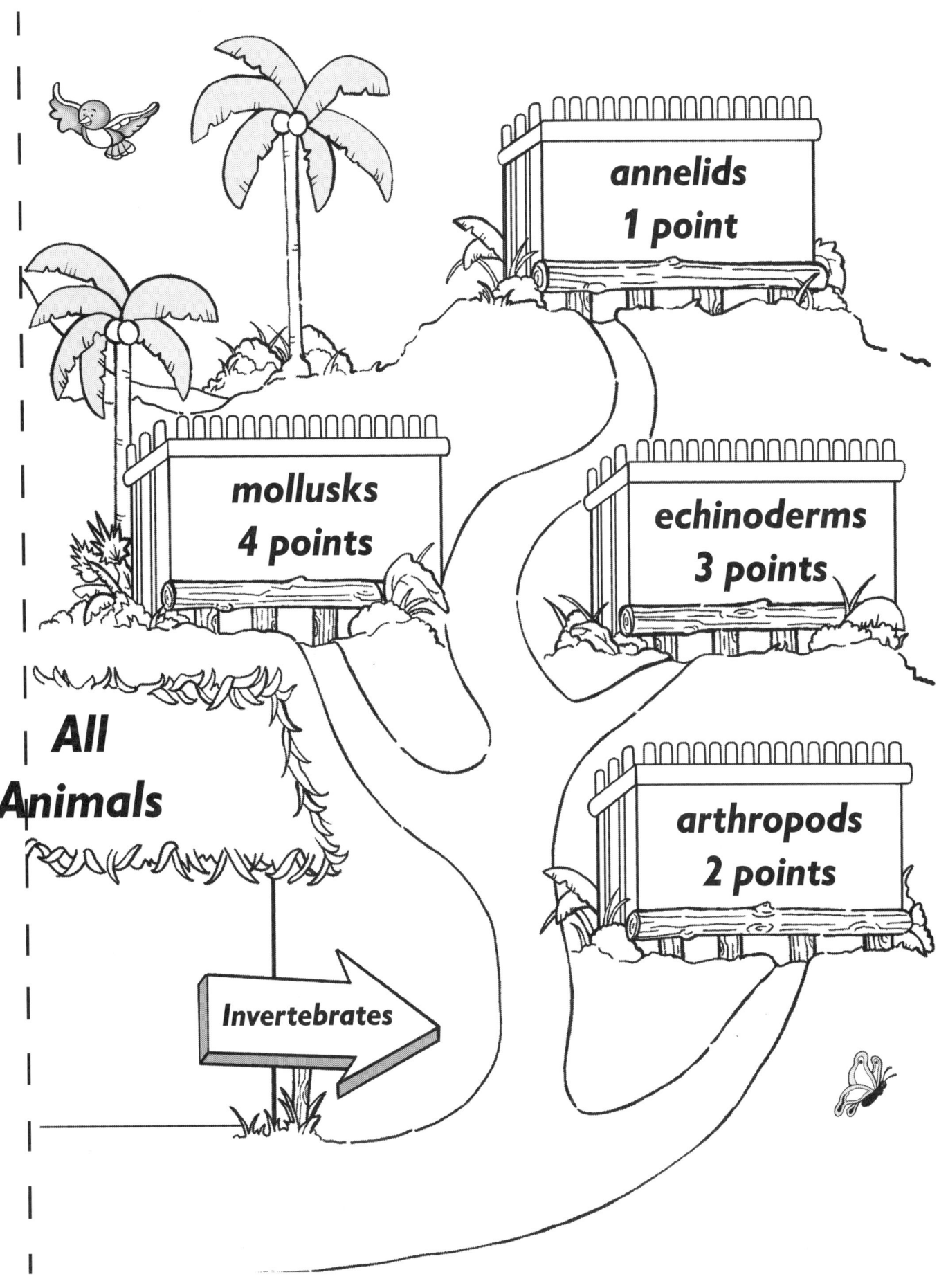

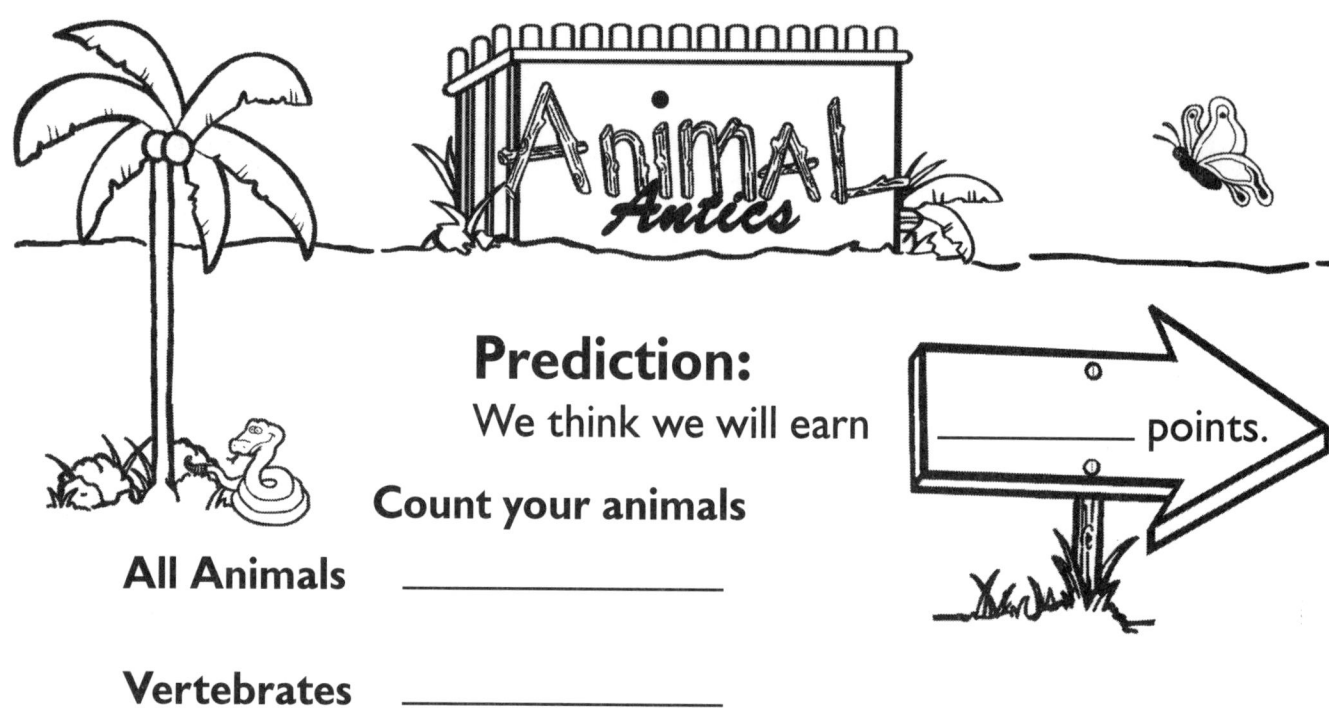

Animal Antics

Prediction:
We think we will earn _____ points.

Count your animals

All Animals _____

Vertebrates _____

Invertebrates _____

Kind of Animal	# of Animals	Points Earned
mammals	_____ × 1 =	_____
fish	_____ × 3 =	_____
birds	_____ × 2 =	_____
reptiles	_____ × 4 =	_____
amphibians	_____ × 5 =	_____
annelids	_____ × 1 =	_____
mollusks	_____ × 4 =	_____
arthropods	_____ × 2 =	_____
echinoderms	_____ × 3 =	_____

Team Total _____

CRITTERS   87   © 2004 AIMS Education Foundation

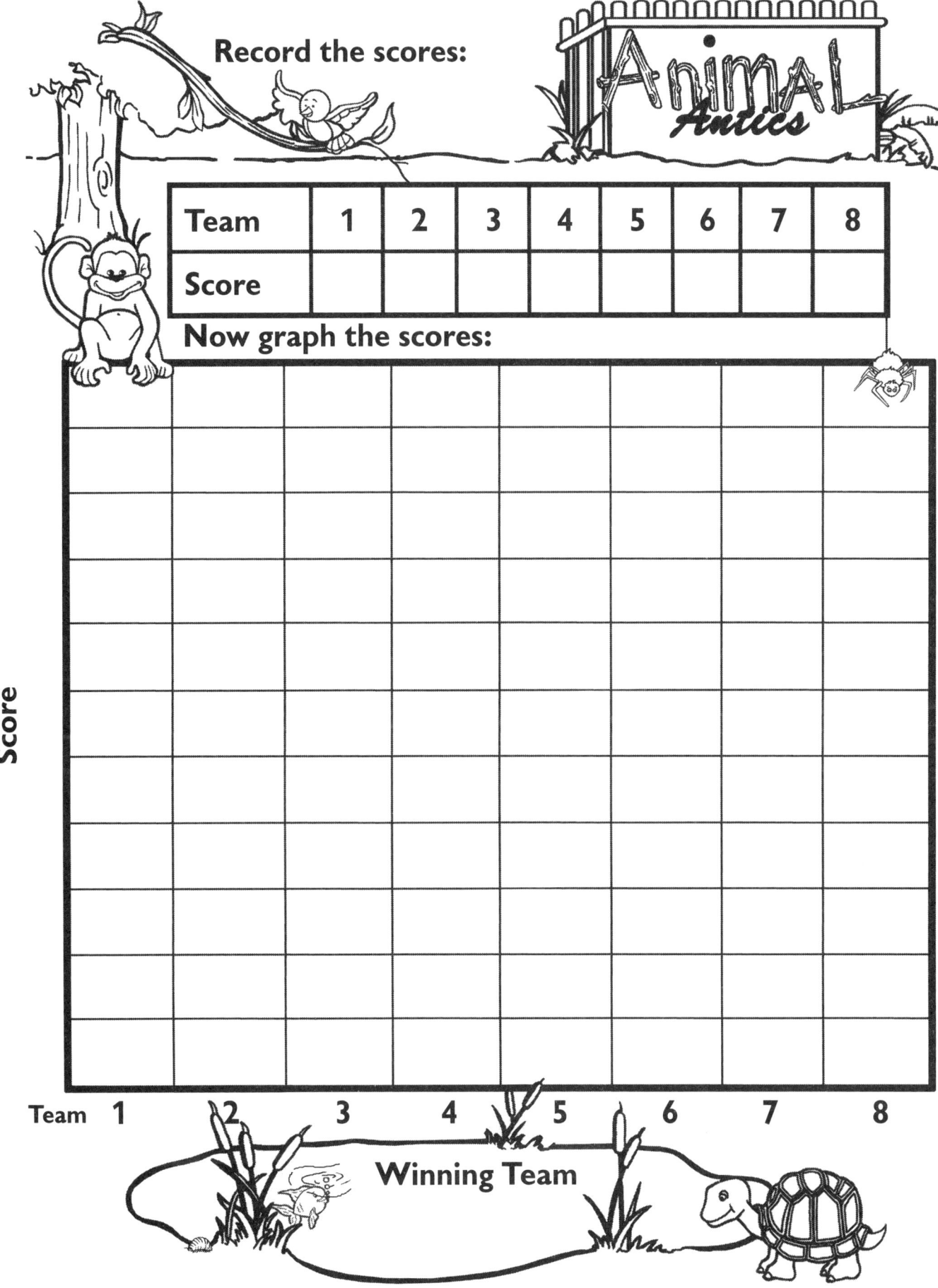

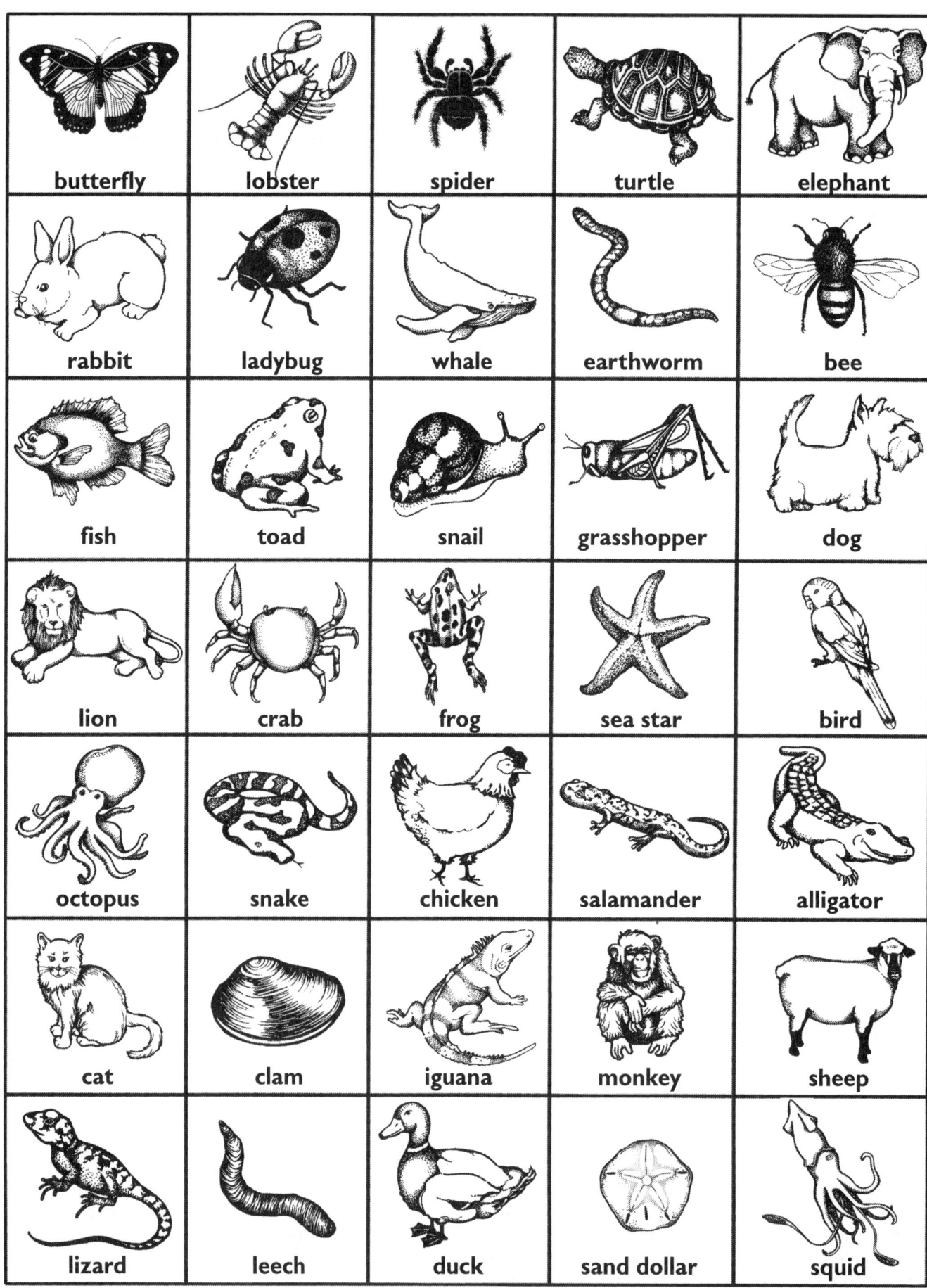

Topic
Fish anatomy

Key Question
How does a fish swim and breathe under water?

Learning Goal
Students will construct a goldfish model that shows all its fins and simulates its breathing.

Guiding Documents
Project 2061 Benchmarks
- *A model of something is different from the real thing but can be used to learn something about the real thing.*
- *Different plants and animals have external features that help them thrive in different kinds of places.*
- *Most living things need water, food, and air.*

NRC Standard
- *Each plant or animal has different structures that serve different functions in growth, survival, and reproduction. For example, humans have distinct body structures for walking, holding, seeing, and talking.*

Science
Life science
    animals
        fish anatomy

Integrated Processes
Observing
Classifying
Comparing and contrasting

Materials
Live goldfish for observation
Scissors
Tape or glue
Student pages

Background Information
Fish breathe by taking dissolved oxygen from water. Fish gulp water through their mouths and pump it over their gills. Most fish have four pairs of gills covered by the operculum. The gills absorb oxygen from the water and replace it with carbon dioxide. If you have a fishbowl without an air bubbler, it is important to change the water every week or two. Otherwise, the fish will use up all of the oxygen in the water.

Fins are movable structures that help a fish swim and keep its balance. The fins found on the top, underside, and tail (the dorsal, anal, and caudal) help the fish remain upright. The caudal fin propels the fish through the water. The pectoral and pelvic fins are located just behind the gills (with the pectoral fins on top) and are used to turn and stop.

The lateral line runs the length of the body and plays an important role for fish as a sensing organ. Fish have no external ears. The lateral line, which has delicate nerve endings all along its length, senses vibrations and movement in the water, thus allowing the fish to "hear."

Management
1. Before beginning the activity, use the activity sheet to make a working model of the fish for demonstration purposes.
2. To assemble the fish, cut out the pieces and tape or glue the fins into their proper positions. Write "H_2O CO_2" across the back of the water strip to indicate the mixture that leaves the gills. Make a cut along each of the dashed lines. Insert the water strip into the mouth so that the side reading "H_2O O_2" is facing up. Give the strip a half twist, and put it through the slit in the gills so that it comes out reading "H_2O CO_2."

Procedure
1. Have students observe a live goldfish for five to 10 minutes. Ask them to locate the various fins.
2. Discuss the function of each fin (see *Background Information).*
3. Ask how fish are able to breathe under water.
4. Accept answers and elaborate by using an assembled model as a visual aid.
5. Hand out the activity sheet and have students cut out the pieces for the fish model. Have them write "H_2O" (water) and "CO_2" (carbon dioxide) on the back side of the water strip. This side represents the water composition after it goes through the gills.
6. Have students assemble their own fish models by gluing or taping the fins and lateral line in the appropriate locations.

CRITTERS   © 2004 AIMS Education Foundation

7. Assist students in making an incision in the paper at the mouth and at one of the gills. Show them how to insert the water pull strip into the mouth of the fish with the side reading "H_2O O_2" facing up. Tell them to give the paper a half twist before pulling it through the gills.
8. Discuss again how the strip models how a fish breathes.
9. If desired, have students get into groups and complete the fish puzzle.

Connecting Learning
1. How does a fish breathe? [It takes in water through its mouth and absorbs the oxygen in its gills.]
2. How is this form of breathing different from our own?
3. Can you think of any other animals that breathe with gills? [tadpoles, isopods]
4. How are these animals similar to fish? How are they different?
5. Why do you have to change your goldfish's water if you have no air pump in your fish bowl? [The fish will use up all of the oxygen in the water and won't be able to breathe.]
6. What are you wondering now?

Extensions
1. Research freshwater and saltwater fish for similarities and differences.
2. Observe isopods or tadpoles to see how they use their gills.
3. Compare and contrast the fins of fish with the fins of aquatic mammals such as dolphins and whales.

Curriculum Correlation
Literature
Parker, Steve. *Eyewitness Books: Fish.* DK Publishing. New York. 2000.

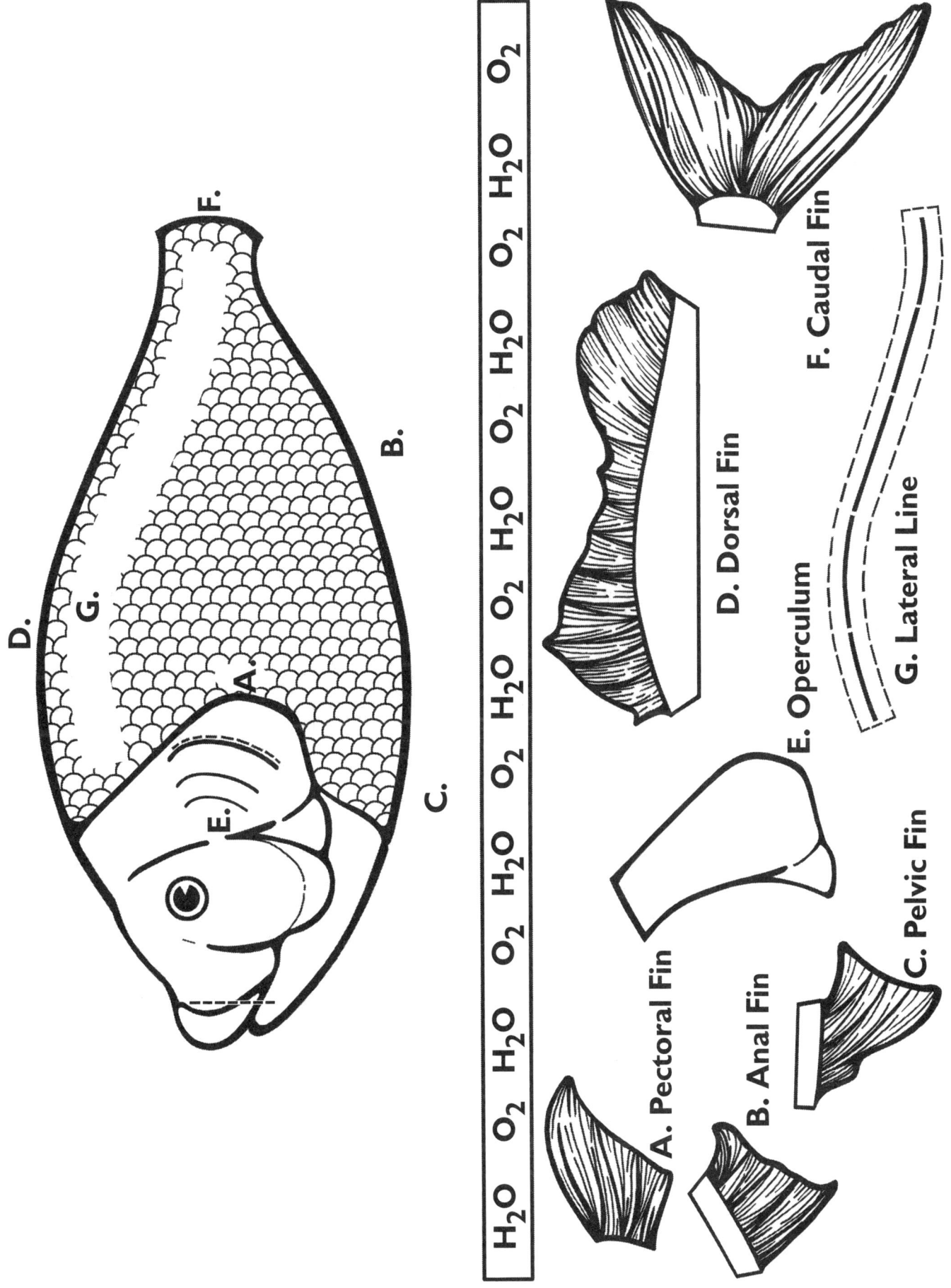

Fishing for Fins

Fish Puzzle

Directions
Randomly place the pieces of the fish puzzle print side up. Each member of the group of four takes a turn placing one piece in position. Each member of the group can move only one piece at a time. A piece that one person has put in place may be moved by another player. This is counted as a move. Talking is not allowed until the puzzle is complete. Play progresses in a clockwise manner. A player may pass if he or she cannot put a piece in place.

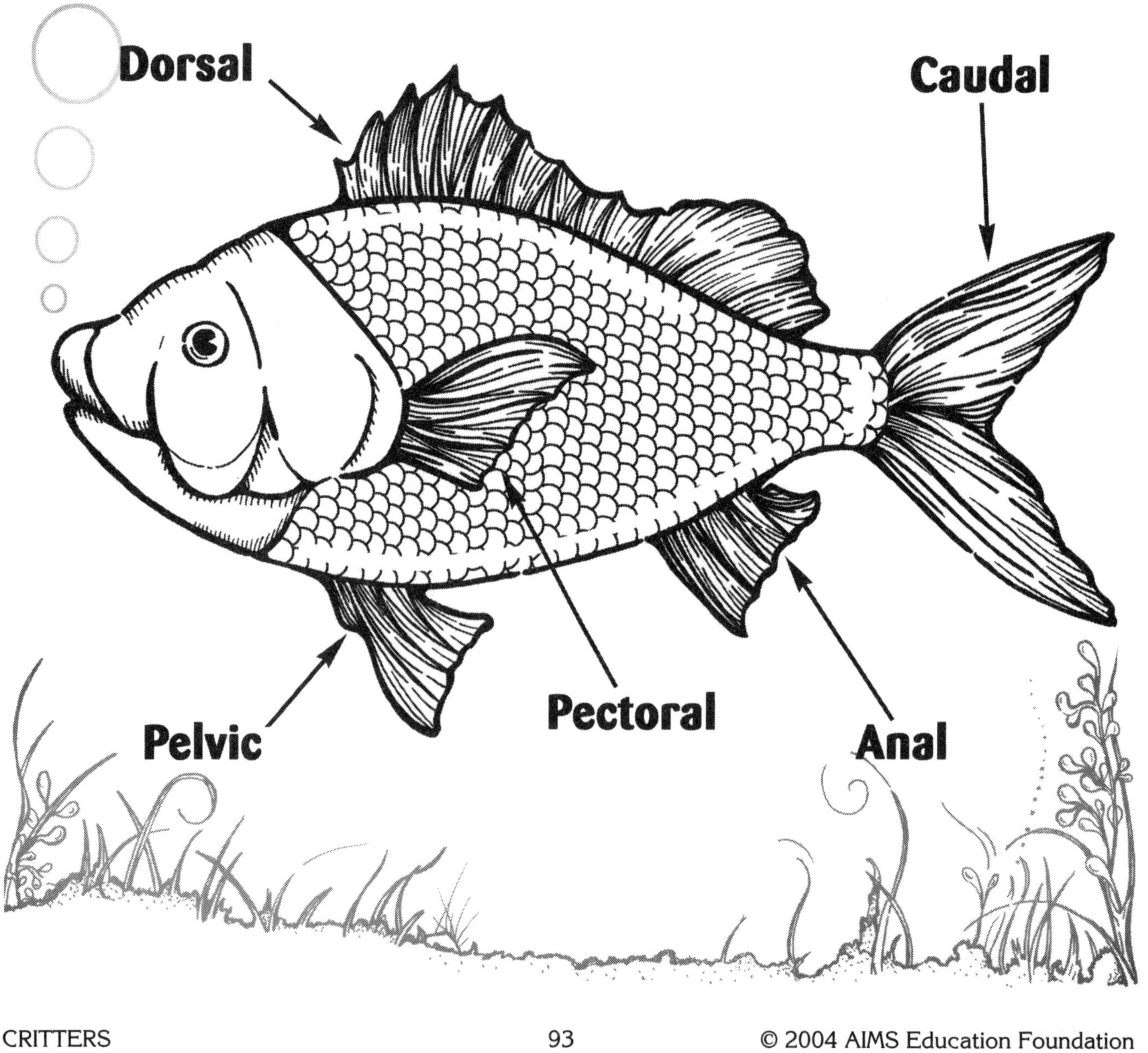

CRITTERS · © 2004 AIMS Education Foundation

Beetle Mania

Topic
Insects, three body parts

Challenge
Put the correct insect body parts together to form each of the six beetles.

Learning Goal
Students will match the descriptions of different insect body parts (head, thorax, and abdomen) to form pictures of actual beetles.

Guiding Documents
Project 2061 Benchmark
- A great variety of kinds of living things can be sorted into groups in many ways using various features to decide which things belong to which group.

NRC Standard
- Each plant or animal has different structures that serve different functions in growth, survival, and reproduction. For example, humans have distinct body structures for walking, holding, seeing, and talking.

Math
Relative sizes
Positional terms

Science
Life science
  insects
    major body parts

Integrated Processes
Observing
Predicting
Comparing and contrasting
Generalizing

Materials
Part One
For each group:
  pictures of insect body parts
  Beetle Mania Clues
  stapler
  paper

Part Two
For each group:
  crayons
  Beetle Mania Cards

Background Information
There are approximately 300,000 different kinds of beetles in the world. Some are considered pests because they eat crops (including trees) and stored grains. Others are considered beneficial because they eat the crop-eating pests.

Like other insects, the adult beetle has three main body parts. These body parts are the focus of this lesson:

Head which includes the antennae, two eyes, jaws, and palps ("feelers" for touching or tasting);
Thorax which is the place where the legs (three pairs) and wings (usually two pairs) are attached; and the
Abdomen which contains the digestive, reproductive, and respiratory systems. (Beetles breathe through holes in their abdomens.)

This activity provides students with illustrations of six heads, six thoraxes, and six abdomens of beetles. When properly arranged, the parts form pictures of six actual beetles.
(Note: The entire thorax is not visible from the top-down view of the beetles in this activity. In these illustrations, the thorax appears much smaller in proportion than it actually is. In order to see the complete thorax, the beetles would need to be viewed from the underside with attention being given to the middle section to which the legs are attached.)

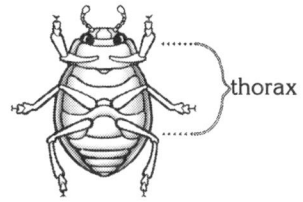

Management
1. This activity is divided into two parts. *Part One* has students follow clues about the bodies of six different beetles. Looking at six options each for heads, thoraxes, and abdomens, students try to determine the correct combinations. In *Part Two*,

they can match their predicted combinations with the actual combinations and then realistically color their beetles following directions on the *Beetle Mania Cards*. (The coloration varies with the species. The coloring schemes suggested on the *Beetle Mania Cards* represent only one species of each featured beetle.)
2. The activity can be done individually or in small groups. If done in groups, one student in each group can read the clues for combining the beetle's body parts while the others in the group match and record the numbers of the body parts.
3. Students should have some prior experience in identifying the body parts of insects. The purposes of this activity are to reinforce the identification of the three body parts and to sharpen observation skills.
4. For younger students, this activity can be done with only three beetles (Potato beetle, Tiger beetle, and Bark beetle). Provide only the body parts with numbers 1, 2, and 3 (Beetles A, B, and C) on them. Older students can use all six beetles.

Procedure
Part One
1. Invite students to share what they know about beetles—common names, where they are found, what they eat, etc. List their responses on the chalkboard.
2. Distribute the pages of "mixed-up beetle parts." Have the students arrange the pages 1-6 with 1 on top (or for younger students, 1-3). Give them a sheet of plain paper and have them place this sheet at the back of the set of pages to make a back cover page.
3. Have students place several (7-8) staples along the left border line of the pages.
4. Direct them to **carefully** cut along the broken lines separating the *Heads* and *Thoraxes* and along the lines separating the *Thoraxes* and *Abdomens*. Caution them to **NOT** cut past the left border line!
5. Tell the students that they will try to find the head, thorax, and abdomen that fits the clues that are read. Once they decide upon a combination, they will record the numbers and later check their answers with the actual number key found in *Beetle Mania Cards*.
6. Read the clues and allow time for the students to match and record the numbers of the predicted body parts.

Part Two
1. After students have matched and recorded the numbers for all six beetles, distribute the *Beetle Mania Cards*.
2. Allow time for students to check their answers and to make any changes necessary.
3. Invite them to color the beetles according to the color key. They can cut off the left margin of the pages, arrange, and glue the colored beetles on pieces of paper.

Connecting Learning
1. Which beetle do you like best? Why?
2. What do all these beetles have in common?
3. On which body part do you find the eyes? ...the antennae? ...the legs?
4. Now that you have observed these beetles, what additional questions do you have about beetles?
5. Where do you think you could find information about beetles?

Extensions
1. Use a hand lens or microscope to more carefully observe beetles.
2. Have students do research on beetles and add to their body-parts book. Ask them to write descriptions of the beetles they are adding.
3. Have students classify the beetles using various attributes such as color, stripes, spots, etc. Using information obtained from research, have them group the beetles according to whether they are considered *harmful* or *beneficial* to plants.
4. Continue the study of insects by investigating protective coloration.
5. Let students choose their own combinations of the different insect body parts and invent names and descriptions for the beetles they make.

Curriculum Correlation
Imes, Rick. *The Practical Entomologist*. Simon & Schuster, Inc. New York. 1992.

Johnson, Sylvia A. *Beetles*. Lerner Publications. Minneapolis, MN. 1982.

Llewellyn, Claire. *Beetles (Minibeasts)*. Franklin Watts, Inc. Danbury, CT. 2002.

Mound, Laurence. *Eyewitness Books: Insect*. DK Publishing. New York. 2000.

Mudd, Maria M. *The Beetle*. Intervisual Books Inc. Santa Monica, CA. 2001.

Pascoe, Elaine. *Beetles (Nature Close-Up)*. Blackbirch Marketing. Woodbridge, CT. 2001.

Beetle Mania Clues

Beetle A
Head: Wide, thin head with small mandibles (jaws)
Thorax: Broad thorax with spots
Abdomen: Roundish abdomen with stripes

My Prediction:
Head _____
Thorax _____
Abdomen _____

Beetle B
Head: Large eyes on side of head, jagged mandibles (jaws)
Thorax: Looks like two gumdrops pushed together
Abdomen: Two diamond shapes and small spots

My Prediction:
Head _____
Thorax _____
Abdomen _____

Beetle C
Head: Small head, very short antennae
Thorax: Line going down the center of its thorax
Abdomen: Long, narrow abdomen with stripes, rounded at base

My Prediction:
Head _____
Thorax _____
Abdomen _____

Beetle D
Head: Large black eyes, no mandibles (jaws)
Thorax: Two large spots, one on each side of thorax
Abdomen: Long, narrow abdomen with stripes

My Prediction:
Head _____
Thorax _____
Abdomen _____

Beetle E
Head: Very long antennae, eyes on the side of its head
Thorax: Plain narrow thorax
Abdomen: Long, abdomen with a few spots

My Prediction:
Head _____
Thorax _____
Abdomen _____

Beetle F
Head: Small and black
Thorax: Broad thorax with a big white spot near the top of each side
Abdomen: Roundish shape with spots

My Prediction:
Head _____
Thorax _____
Abdomen _____

CRITTERS © 2004 AIMS Education Foundation

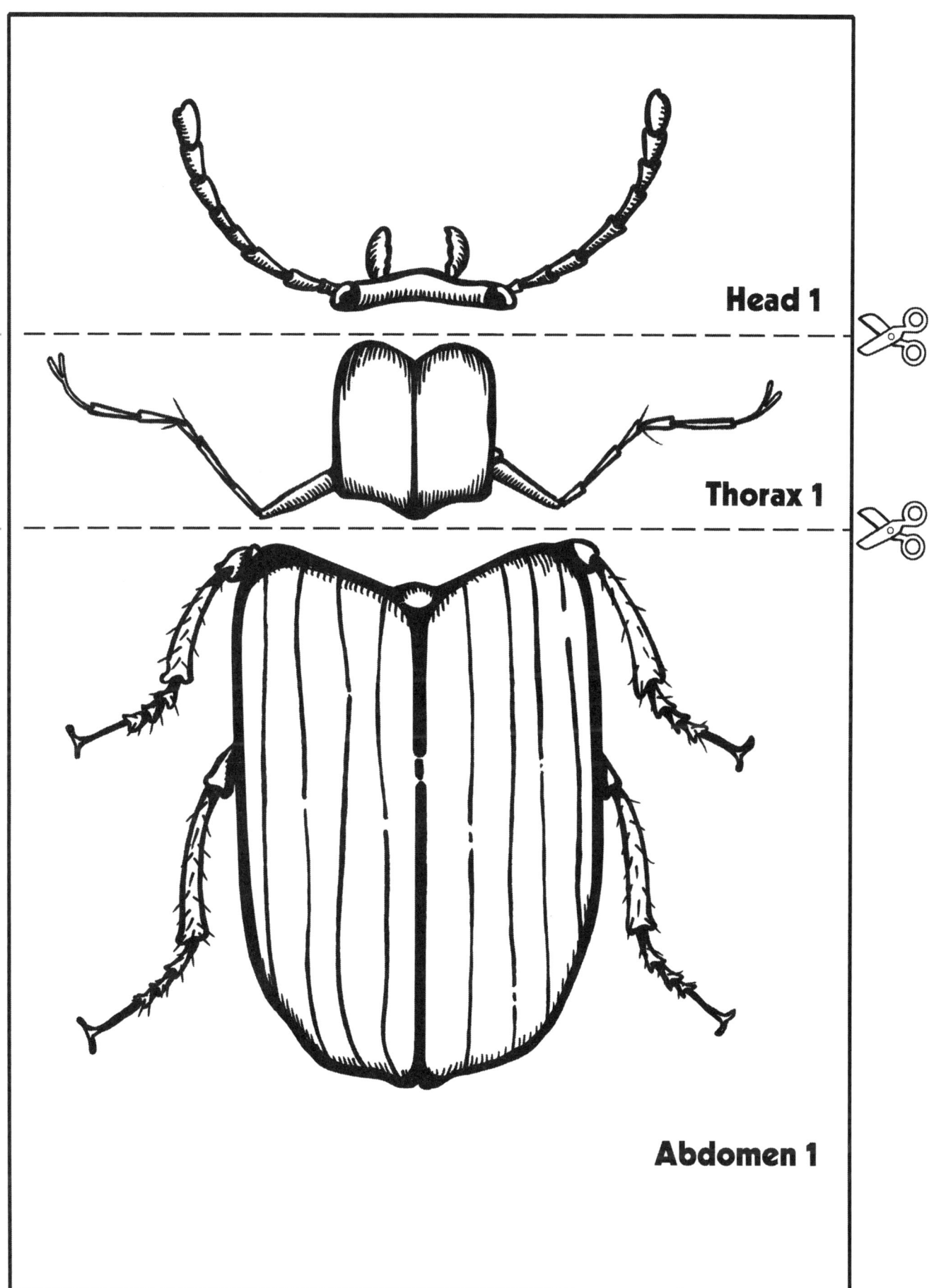

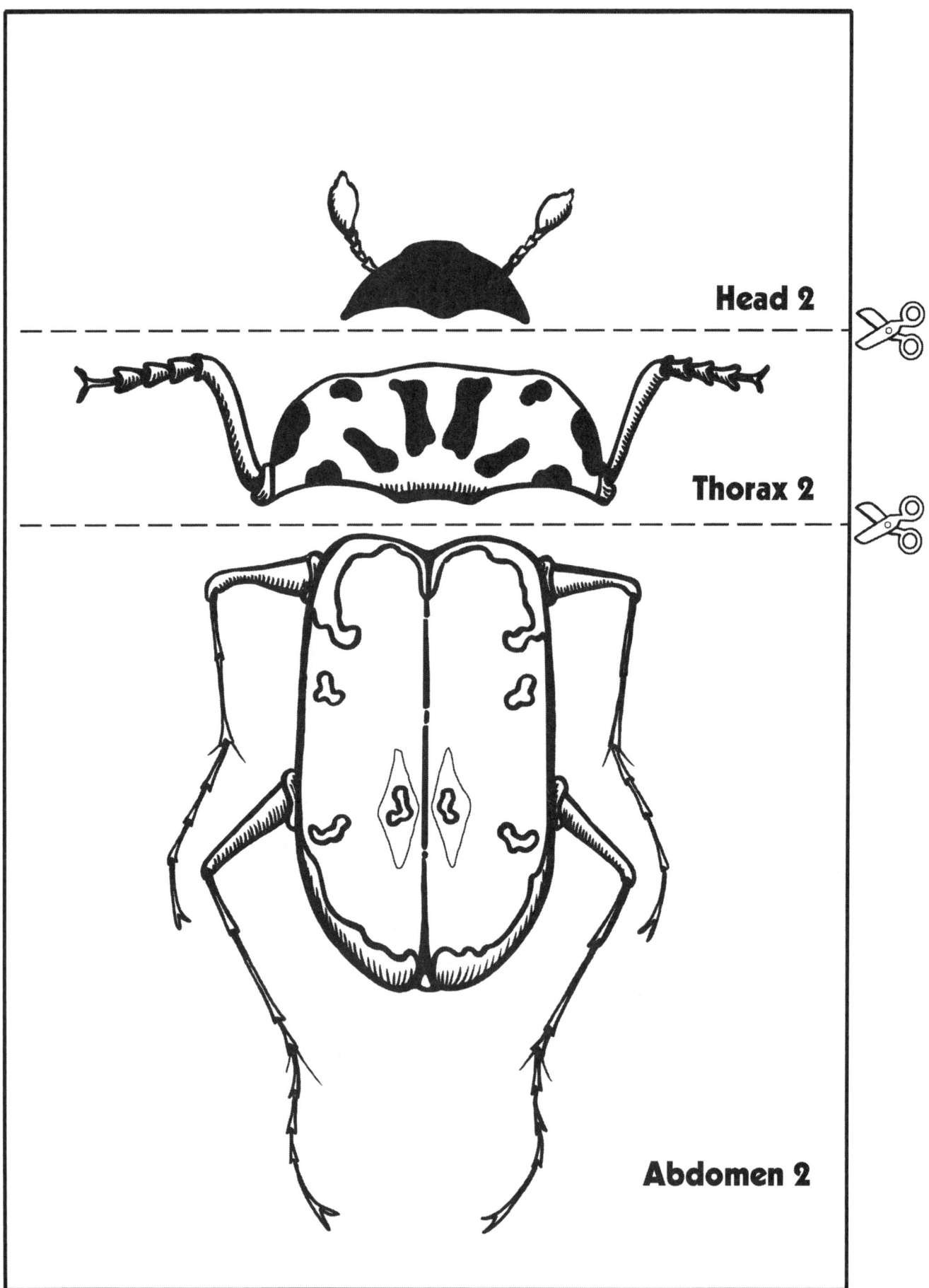

CRITTERS

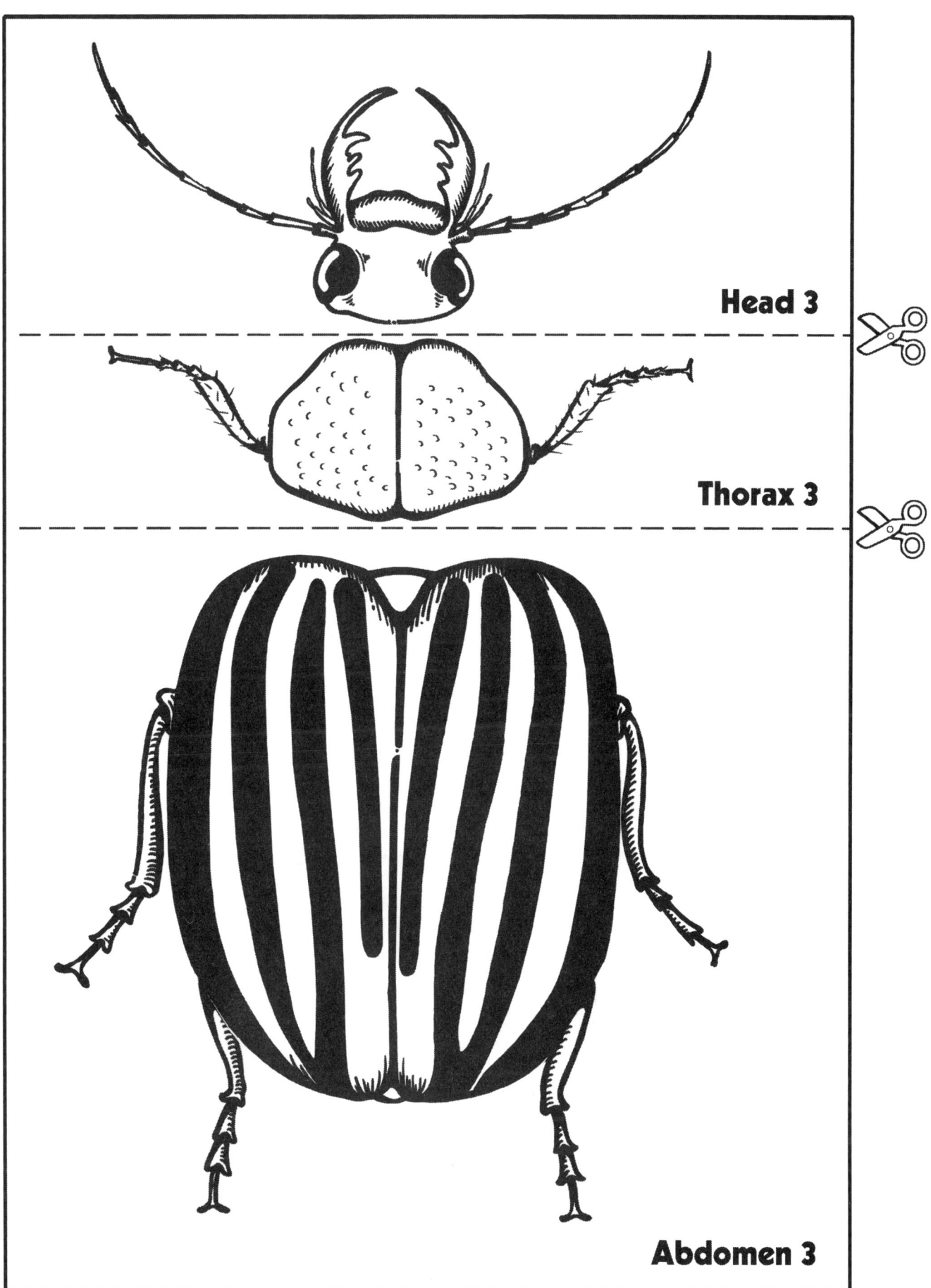

Head 3

Thorax 3

Abdomen 3

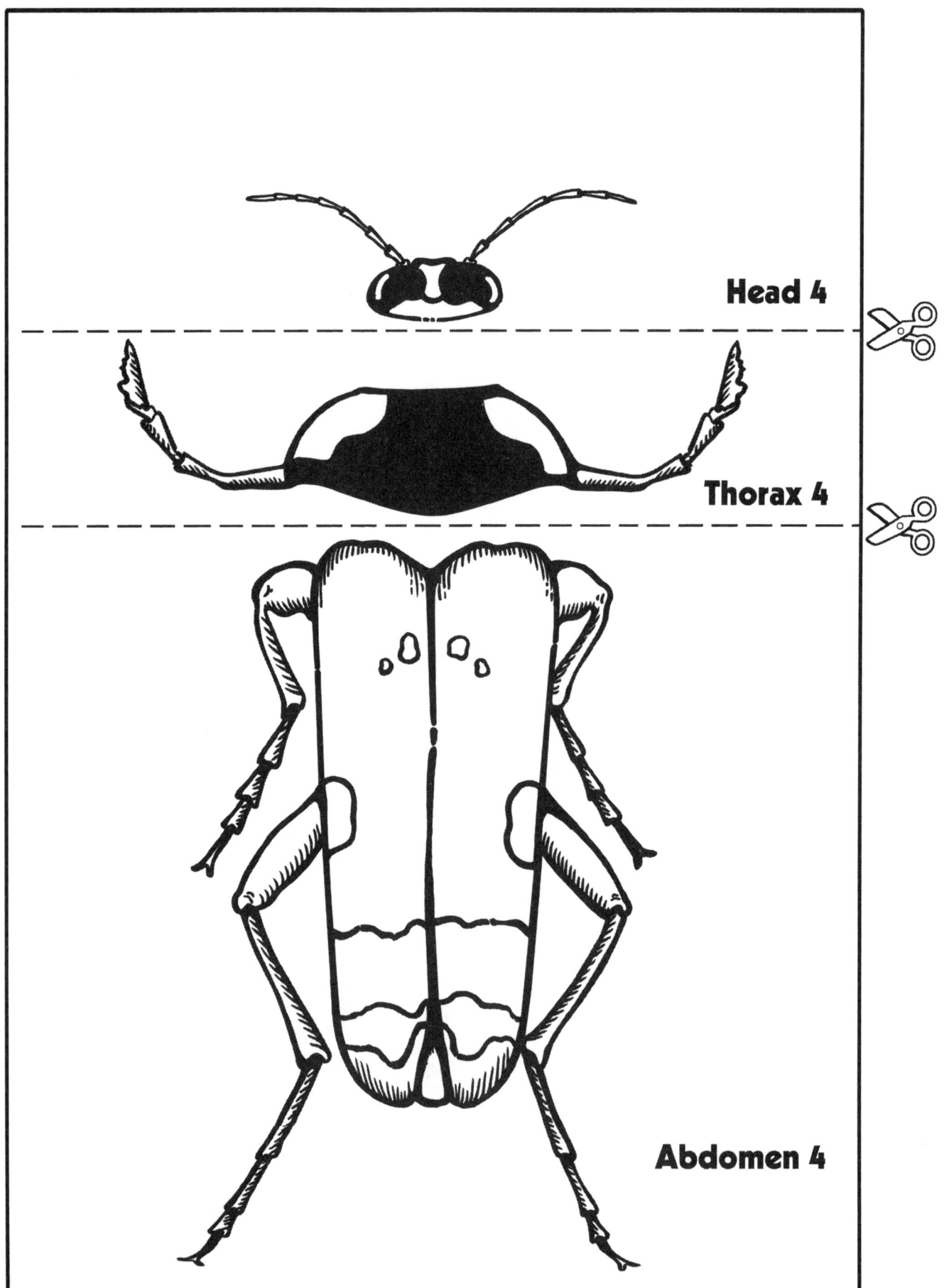

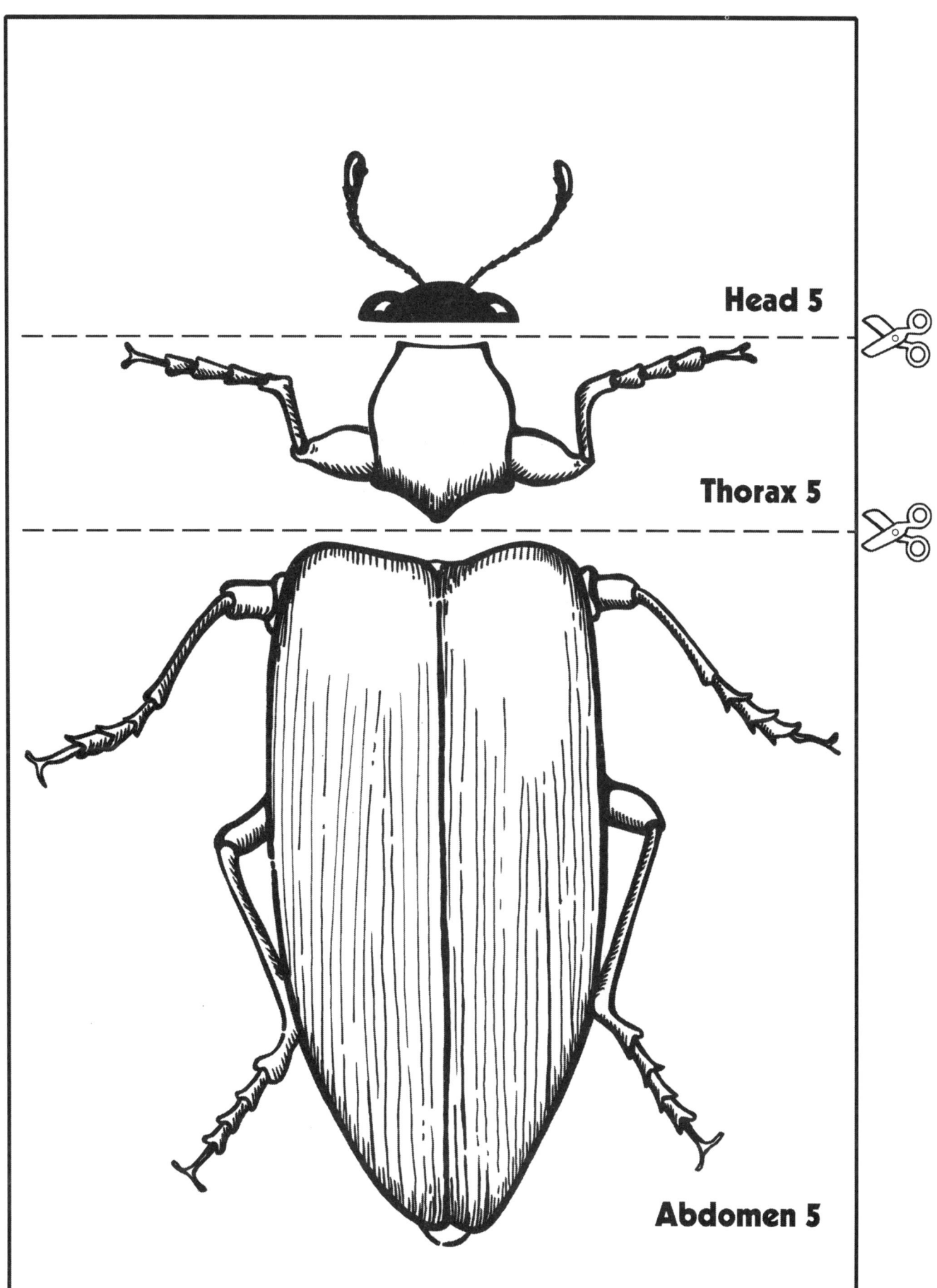

CRITTERS 103 © 2004 AIMS Education Foundation

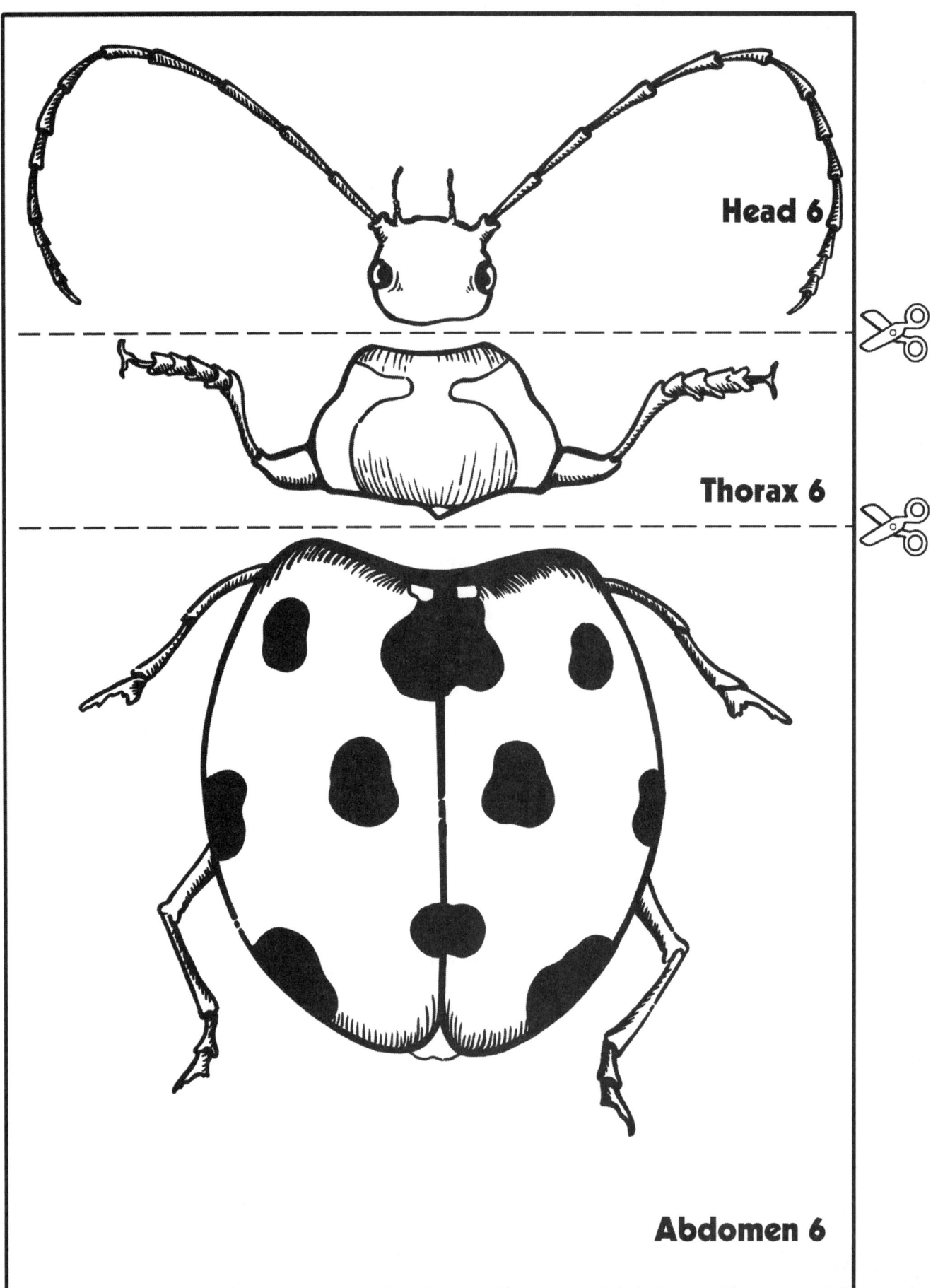

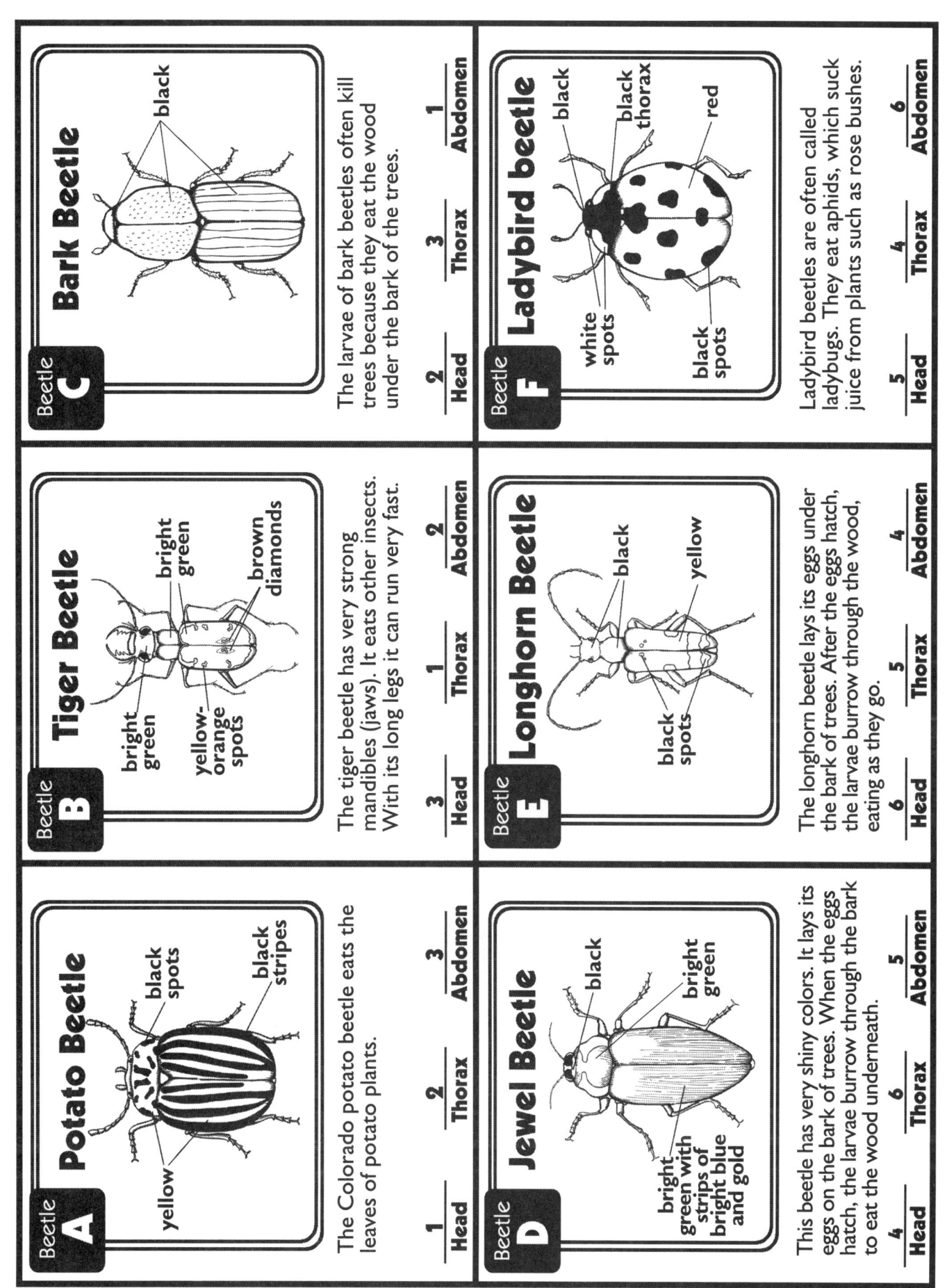

A look at Lepidoptera

Topic
Classification

Key Question
How can we learn to distinguish butterflies and moths by paying attention to their attributes?

Learning Goals
Students will:
1. classify butterflies and moths by sorting by attributes,
2. record their sorts using labeled two-circle and three-circle Venn diagrams, and
3. infer characteristics of butterflies and moths by the results of their sorts.

Guiding Documents
Project 2061 Benchmarks
- A great variety of kinds of living things can be sorted into groups in many ways using various features to decide which things belong to which group.
- Features used for grouping depend on the purpose of the grouping.

NRC Standards
- Use data to construct a reasonable explanation
- Scientists use different kinds of investigations depending on the questions they are trying to answer. Types of investigations include describing objects, events, and organisms; classifying them; and doing a fair test (experimenting).

Math
Sorting
Venn diagrams

Science
Life science
   classification
   diversity

Integrated Processes
Observing
Comparing and contrasting
Classifying
Predicting
Generalizing

Materials
Three grouping circles per group
Student pages
Glue stick
Index cards, 3" x 5"
Pocket chart, optional

Background Information
Butterflies and Moths

Butterflies and moths are insects that belong to the order *Lepidoptera*. There are over 170,000 known species of lepidoptera, with 90 percent of those being moths. Lepidoptera means scaled wings. The wings of the butterflies and moths are covered with tiny overlapping scales that account for the distinctive patterns and coloration. Butterflies and moths are extraordinarily diverse in their sizes, shapes, and coloration. They have adapted to basically every climate: Arctic tundra and mountain summits to the rainforests of the tropics.

All butterflies and moths have three body parts (head, thorax, and abdomen), antennae, six legs, and two pairs of wings (forewings and hindwings).

Characteristics

Butterflies are generally day-flying insects. They normally have thin antennae with thickened tips. Their bodies tend to be thinner and not as "furry." Butterflies have a variety of wing shapes and sizes; however, their wings are generally large and rounded when compared to the wings of moths. Butterflies are easiest to identify when they are at rest. Their wings are held together above their backs.

Moths are so diverse that it is difficult to pinpoint definite attributes. Some characteristics that might be observed are: antennae are often feathered, making them appear much thicker than those of the butterflies; the moths' antennae also do not have the thickened tips; the moths' wings are often elongated; when at rest, the wings fold over the back and look like the peaked roof of a house.

Students will be somewhat limited with the black-and-white copies. They will, however, be able to compare body sizes, wing shapes, antennae differences, etc. The more they sort the lepidoptera, the more characteristics they will notice.

CRITTERS

Venn Diagrams

Venn diagrams are used to represent sets with something in common. They can be used to compare and contrast two or more attributes. If students have not previously used Venn diagrams, begin with one circle. The inside of this circle is for lepidoptera that have a certain attribute and the area surrounding the circle is for lepidoptera that do not have that attribute.

A two-circle Venn has three areas plus the area outside the circles. If the labels for the circles are *fat body* and *feathery antennae*, one area is for those lepidoptera that have fat bodies, one area is for lepidoptera with feathery antennae, and the intersection is for lepidoptera with both fat bodies and feathery antennae. Again, the area outside the circles is for lepidoptera that do not have either attribute.

The three-circle Venn diagram follows the same pattern. Now, there are seven regions. Each region overlaps each of the others, and the center is an overlap of all three attributes. Lepidoptera within this area will have all three attributes.

Activity

In this activity, students will generate the various attributes that they observe in the pictures of lepidoptera. They will then select two (then three) of these attributes to see which lepidoptera share these attributes. Through the process of many sortings, students will begin to see that there are characteristics that help distinguish butterflies from moths. Students should be able to make generalizations such as, "Most moths have feathery antennae and fat bodies."

Management
1. Have students work in groups of three or four.
2. The illustrations on the first five student pages are approximately life-size. The sixth student page is to be used as cutouts for students to record their sortings on the Venn diagram pages and for optional extensions. Each group will need two cutout pages.
3. Each group will need a set of the to-scale illustrations. Do not hand out the illustrations that have numbers and names until students are ready to record their sortings on the Venn diagrams.
4. Students can use index cards for labels when they are using their grouping circles with the to-scale illustrations. The index cards can also be used for *Extension 1*.
5. The table at the end of this text is intended for teacher information. Students should be encouraged to do some research to locate information about the various butterflies and moths.

Procedure
1. State the *Key Question* and the first *Learning Goal*.
2. Distribute the first five student pages to each group. Ask students to cut out the boxes that enclose the butterflies and moths.
3. Direct them as a group to write down at least 10 observations they can make from the illustrations.
4. As a whole class, solicit one observation from each group, continuing until all observations have been shared. As students are sharing, make sure that there is a common vocabulary that is being developed and that the attributes are clearly understood. To encourage careful listening, ask that no observations be repeated. It may be helpful to list these observations on chart paper for future reference.
5. Distribute the grouping circles to each group. Tell them to choose one and to place it where all members of their group can easily see it.
6. Have a student select an attribute from the list that was shared. Direct the groups to place all the illustrations that contain that attribute into the circle. Check to see if there is agreement among all groups.
7. Have another student select another attribute and follow the same procedure.
8. Direct students to place another grouping circle next to the original one. Now have students select two different attributes, placing the illustrations in the respective circles. Discuss the need to overlap circles when some illustrations show both attributes.
9. Distribute index cards and have students make labels of two attributes chosen by their group. Direct them to place the labels in the respective circles. Give students time to sort their illustrations into the grouping circles.
10. After the group has come to a consensus, ask students to remove their illustrations from the grouping circles. As soon as all groups are ready, have the groups rotate clockwise to at least three other groups to perform the sorting of illustrations by the attributes written on the index cards.
11. Have students return to their original area. Inform them of the second *Learning Goal*. Have them add a third grouping circle to their set. Using an overhead transparency, discuss the various regions. Add labels and discuss the overlapping regions.
12. Ask students to choose attributes, make labels, and sort their illustrations.
13. Invite students to share their discoveries.
14. Ask them if they have found any two attributes that are shared by many of the illustrations. If so, ask what those attributes are. Through the process, lead students to see that many illustrations share common characteristics and can be used to help identify these lepidoptera as butterflies or moths.

15. State the third *Learning Goal.* Ask students to label their circles with *fuzzy antennae, fat bodies,* and *elongated wings.* Have them sort their illustrations. Discuss their findings. [There are many illustrations that do not fit any of the categories. There are many illustrations that have all three characteristics and so are found in the very center intersection.]
16. Tell them that these characteristics are common to moths. Ask what characteristics are common to butterflies.
17. Have students predict which illustrations are butterflies and which are moths. Check these predictions with answers from the table.
18. Distribute the Venn diagram pages and the two cutout pages per group. As a group, have students choose attributes and cut and glue the illustrations in the appropriate regions. Encourage them to trade papers with other groups to check for accuracy.

Connecting Learning
1. What were some unusual attributes you observed in the illustrations?
2. What are Venn diagrams? What are Venn diagrams used for?
3. Besides looking at attributes of butterflies and moths, what other uses can you think of for Venn diagrams?
4. How many butterflies did you count? How many moths? Did everyone agree? Why or why not?
5. For which illustrations was it hardest to decide whether they were butterflies or moths? What made it hard to decide?
6. If you were to "design" your own butterfly, what attributes should you include? What if you were to design a moth, which would you include?
7. Where does illustration number 6 go in a Venn diagram labeled *thin antennae, wings with tails,* and *eyespots*?
8. What are you wondering now?

Extensions
1. Let students choose a butterfly or moth that has been featured in this activity. Have them cut out the butterfly/moth label and glue it to one end of an index card. Encourage students to do some research on their chosen lepidoptera and write interesting facts on the lower end of the index card. These cards can be displayed in a pocket chart or on a bulletin board. Students may wish to use another (white) index card and colored pencils to illustrate the coloration of their lepidoptera.
2. Have students pin the cutout pictures of the lepidoptera to a world map in order to see the range of their distribution.
3. Encourage students to investigate other lepidoptera, especially those indigenous to their area. Have them discuss whether they are considered beneficial or destructive. (Make certain that students understand that most often it is when the lepidoptera are in the caterpillar stage that they are considered destructive.)

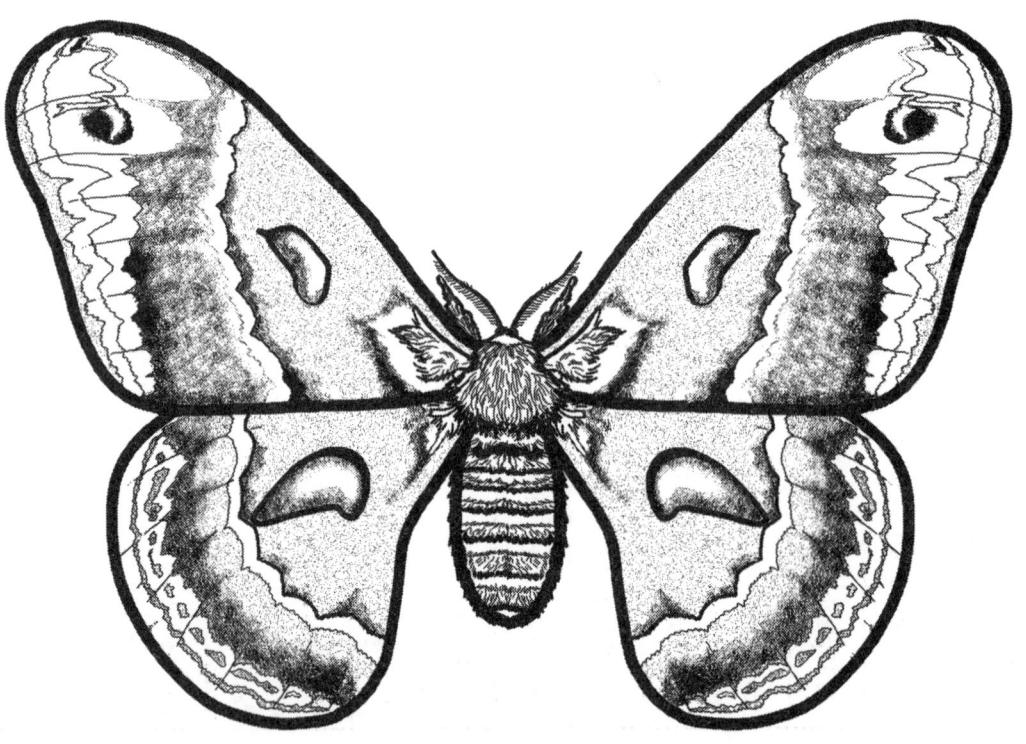

Name	Butterfly/Moth	Characteristics	Distribution	Wingspan
1. Hercules	moth	feathery antennae, tails on hindwing look like feet, outer edges of wings appear scalloped	New Guinea, Australia	16.5-25 cm
2. Cizara Hawkmoth	moth	two small eyespots near base of the forewings, fat body with vase-like design	Australia	5-7 cm
3. Twin-Spotted Sphinx	moth	large furry body with narrow light bands, notched tips on forewings, eyespots on hindwings have black bars	North America	5-8 cm
4. Pipevine Swallowtail	butterfly	very few markings on forewings	North America	7.5-11 cm
5. Peacock	butterfly	two pairs of eyespots, one on forewing and one on hindwing	Europe, Asia	5.5-6 cm
6. Hewiton's Blue Hairstreak	butterfly	curly tails on hindwings, narrow black border around wings	tropical South America to Mexico	4.5-6 cm
7. Old World Swallowtail	butterfly	hindwings have fairly short tails; thin, pale leading edge to forewings	Europe, Asia, Arctic regions	7-10 cm
8. Long-Tailed Skipper	butterfly	has curved tips on antennae, white spots on forewings	North/South America	4-5 cm
9. Acacia Carpenter Moth	moth	lower two-thirds of antennae furry, narrow white rings on abdomen, pattern on forewings looks like scales	Australia	13-20 cm
10. Great Peacock Moth	moth	feathery antennae, elaborate body pattern matches wing pattern	southern Europe, North Africa, Asia	10-15 cm
11. Luna Moth	moth	dark band along leading edge of the forewing, plump furry body, two pairs of eyespots	United States, Mexico	7.5-10.8 cm
12. Buckeye	butterfly	two small bands on each of the leading edges of the forewings, three pairs of eyespots	North America	5-6 cm
13. Hakea Moth	moth	small hook on tip of forewing	Australia	4.5-5.5 cm
14. Cecropia Moth	moth	striped, furry body; crescent-shaped spots on wings	North America	11-15 cm
15. Common Clubtail	butterfly	dark scalloping of hindwings, elongated forewings	India, Malaysia	9-13 cm
16. Pink-Spotted Hawkmoth	moth	small hindwings have ripple-looking bands, large body has light and dark horizontal bands	South/Central America, southern United States	8-12 cm
17. Monarch	butterfly	hindwings are more rounded than forewings, edges of the wings are white spots in black border	Americas, Indonesia, Australia	7.5-10 cm

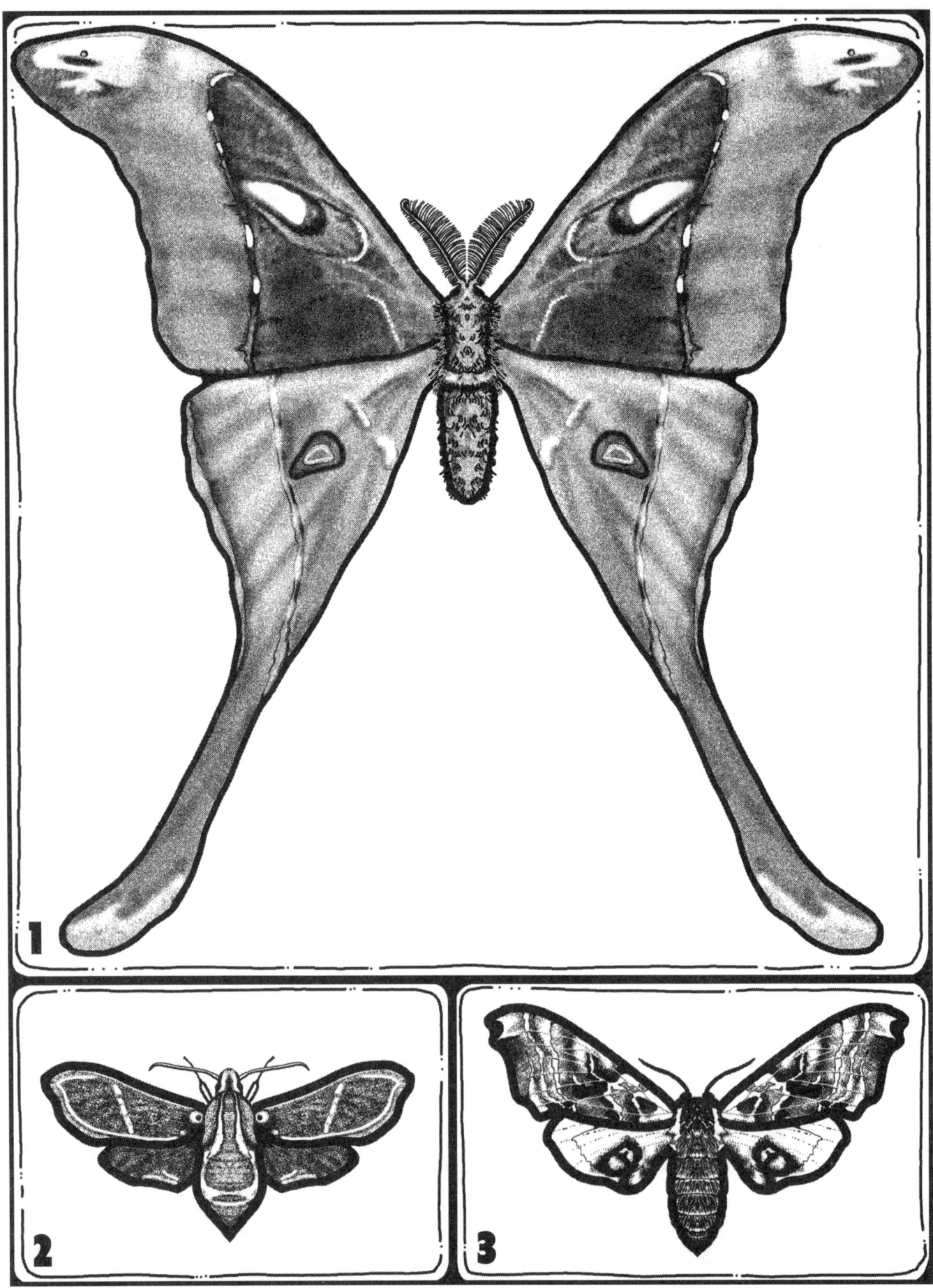

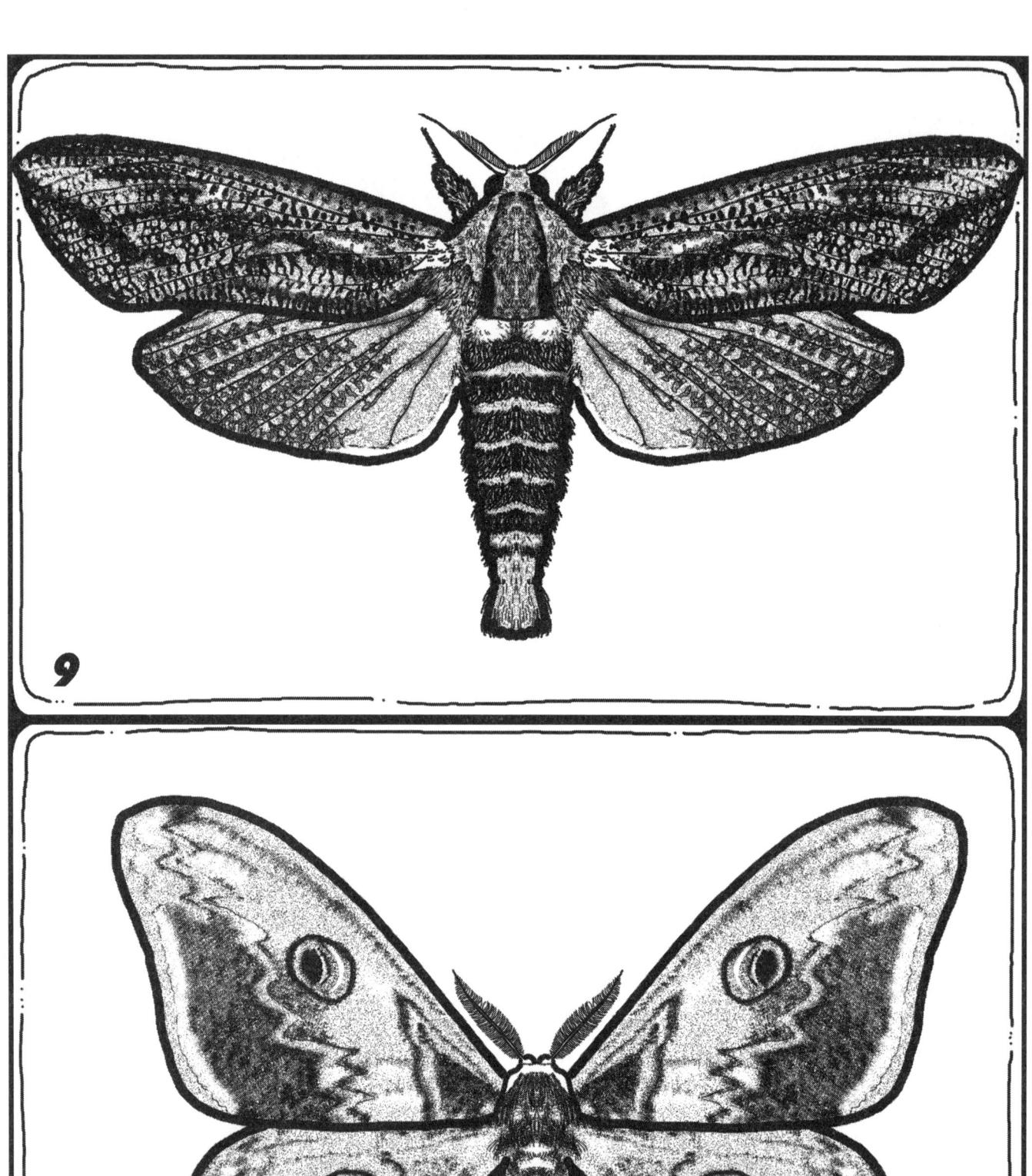

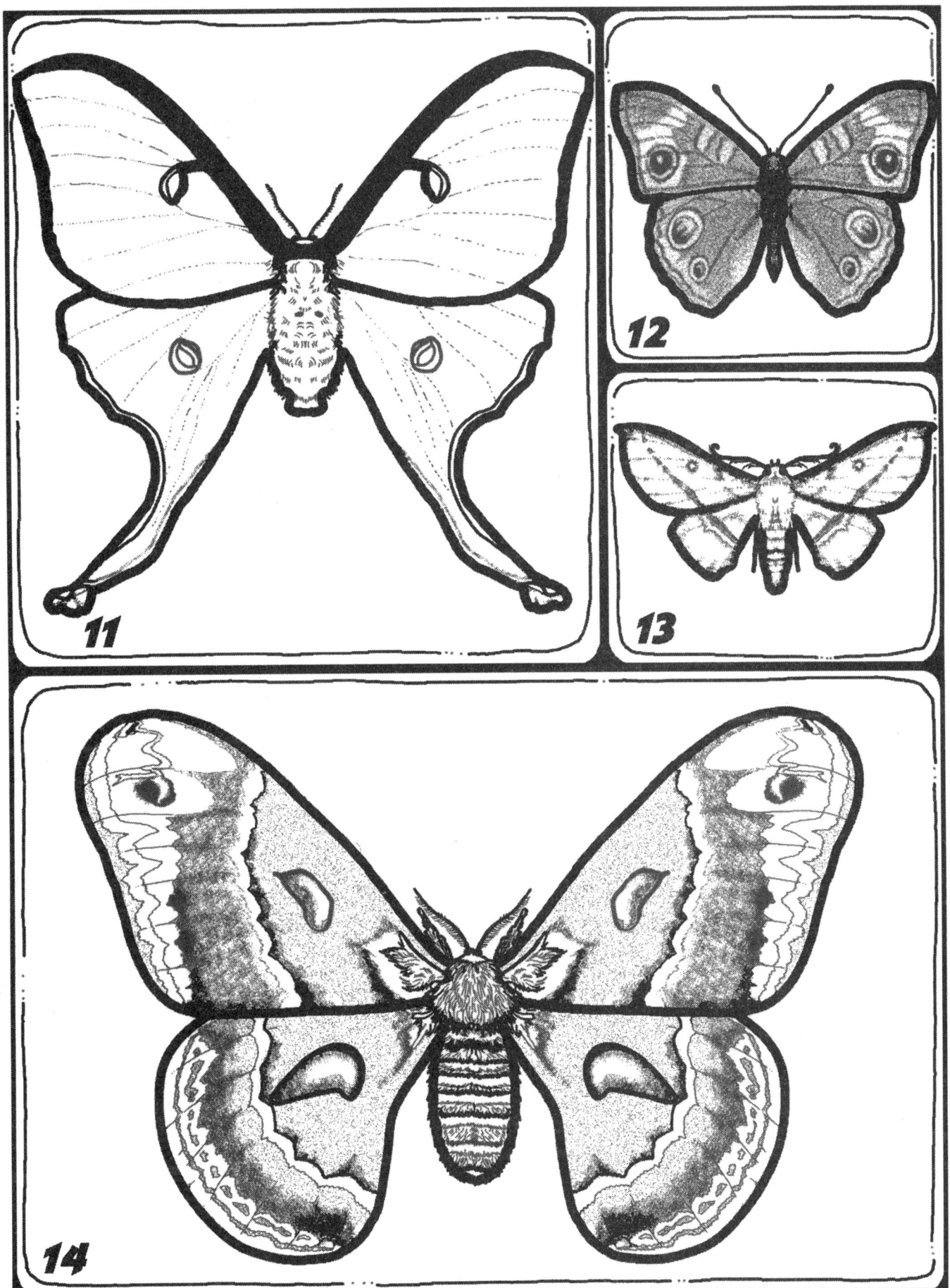

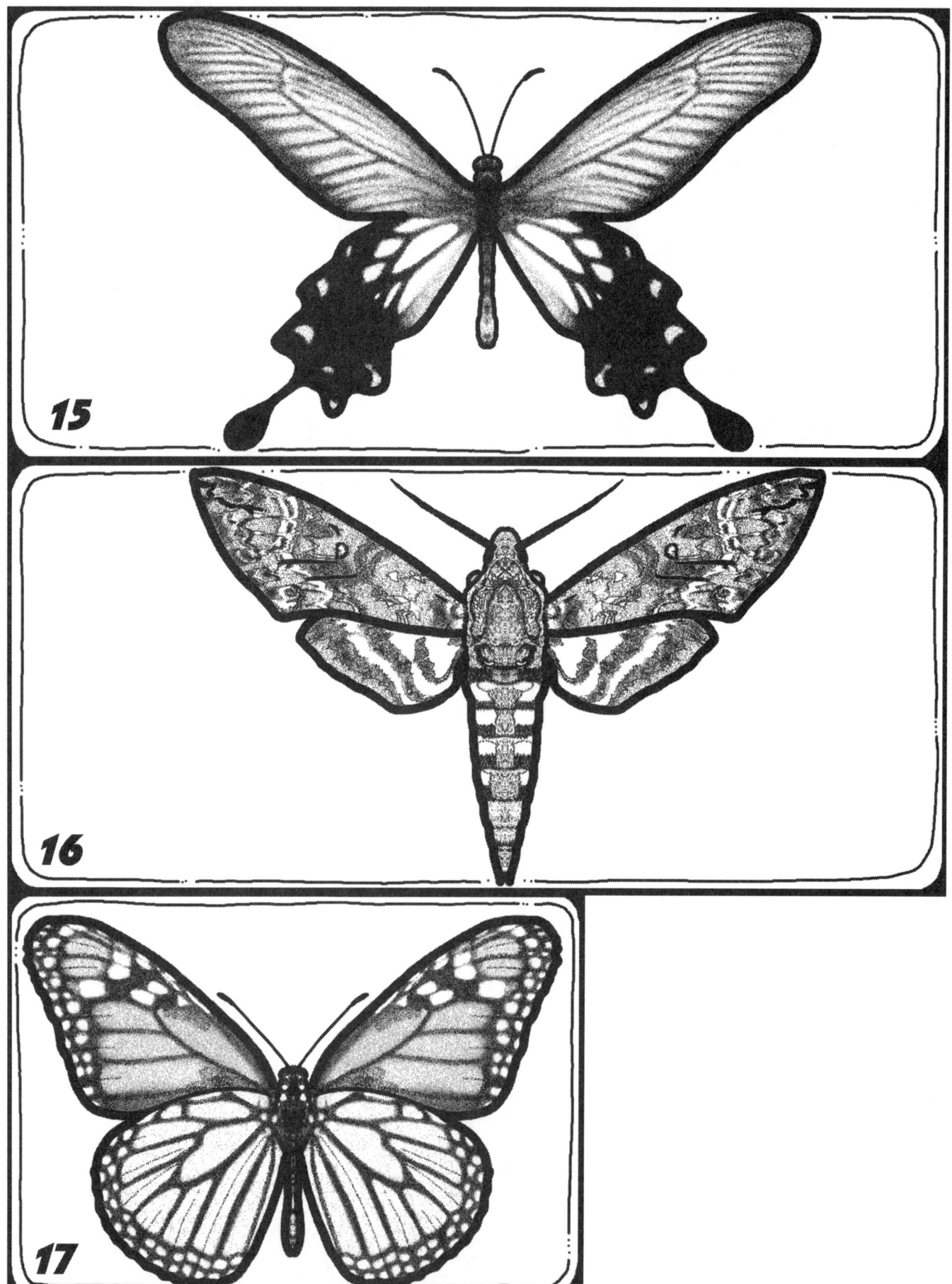

A look at Lepidoptera

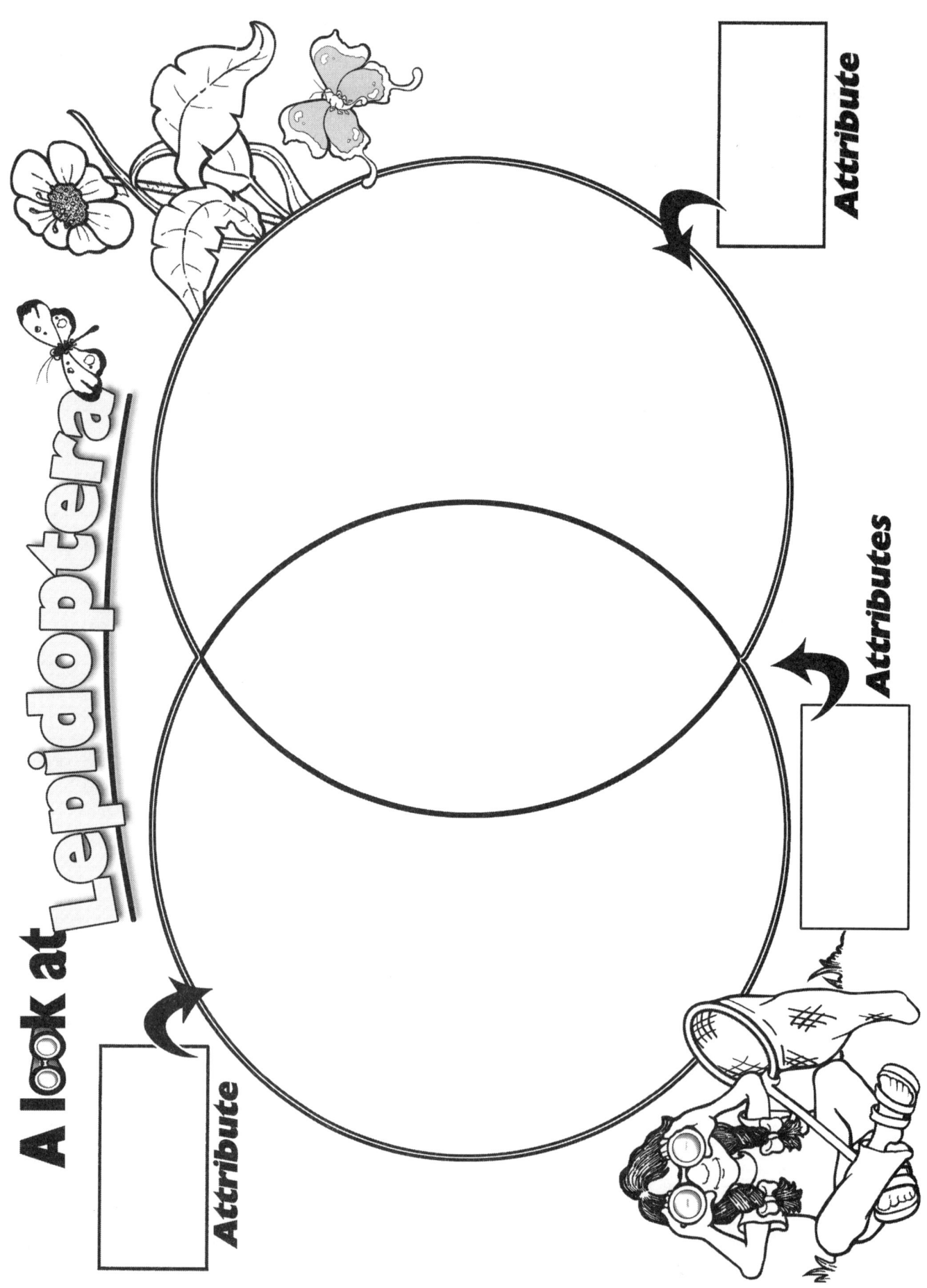

A look at Lepidoptera

Label the circles and place the butterflies and moths in the regions where they belong.

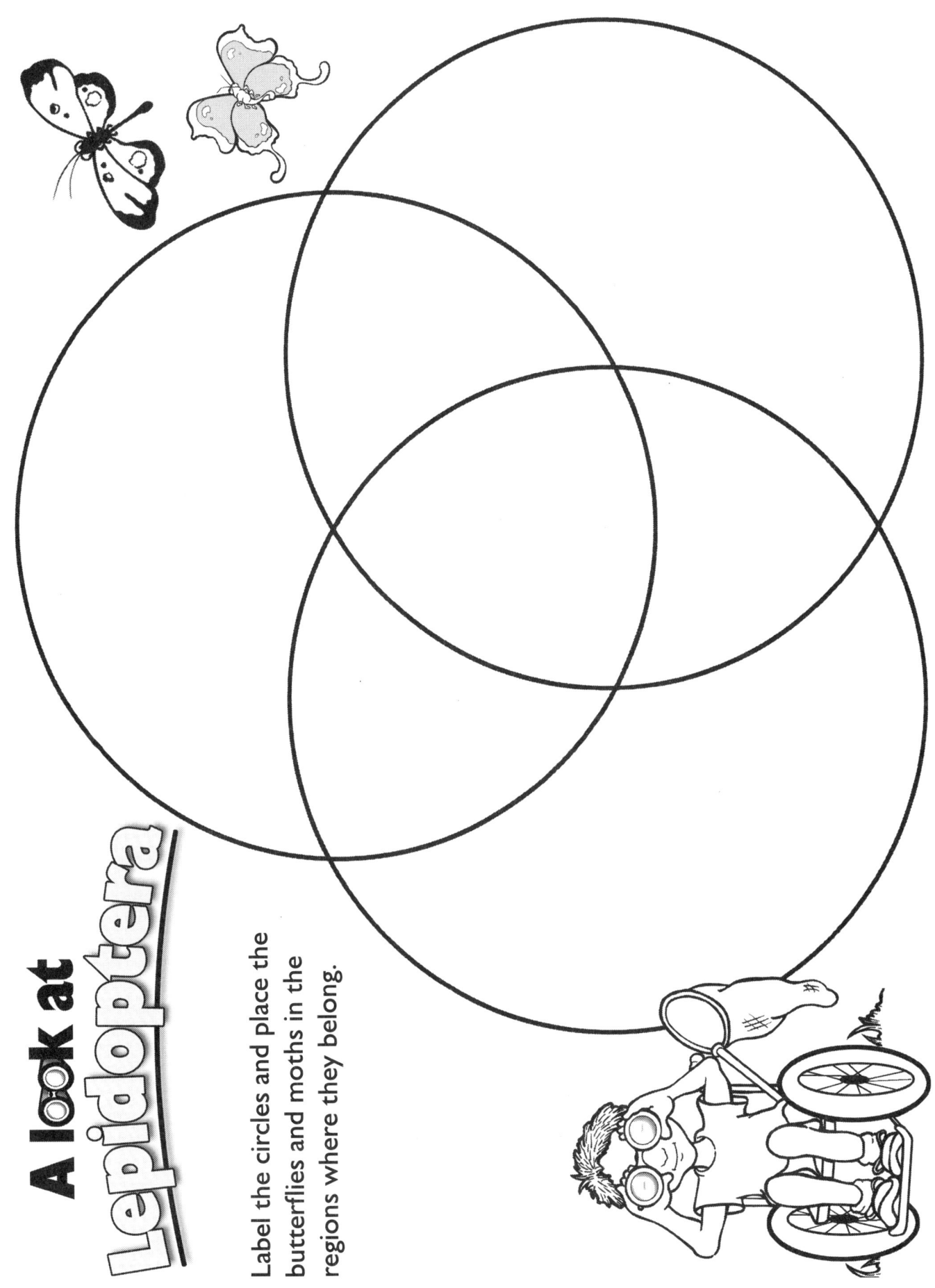

Frog and Toad are Kin

Topic
Characteristics of frogs and toads

Key Questions
What are some characteristics of frogs and toads? How are they alike? How are they different?

Learning Goals
Students will:
1. research characteristics of frogs and toads;
2. determine which characteristics of frogs and toads are the same and which are different; and
3. identify characteristics as belonging to frogs, toads, or both.

Guiding Documents
Project 2061 Benchmarks
- Plants and animals have features that help them live in different environments.
- A great variety of living things can be sorted into groups in many way using various features to decide which things belong to which group.
- Buttress their statements with facts found in books, articles, and databases, and identify the sources used and expect others to do the same.
- Locate information in reference in books, back issues of newspapers and magazines, compact disks, and computer databases.

NRC Standards
- Each plant and animal has different structures that serve different functions in growth, survival, and reproduction. For example, humans have distinct body structures for walking, holding, seeing, and talking.
- Organisms have basic needs. For example, animals need air, water, and food; plants require air, water, nutrients, and light. Organisms can survive only in environments, and distinct environments support the life of different types of organisms.

Science
Life science
 animal characteristics
  frogs and toads

Integrated Processes
Observing
Comparing and contrasting
Collecting and recording data
Analyzing

Materials
Part One
 The Critter Connection: Frogs and Toads
 #19 rubber bands, one per student
 research materials (see *Management 1*)
 chart paper
 colored pencils (see *Management 3*)
 student pages

Part Two
 scissors
 glue sticks
 colored construction paper (see *Management 3*)
 frog and toad pictures (see *Management 4*)

Background Information
Frogs and toads are by far the most numerous and common amphibians on Earth. There are more than 3500 known species of frogs and toads in the world, and more are being discovered every year. There are many common misconceptions about frogs and toads and the differences between them. This activity will give students the opportunity to research some of those differences and reverse many misconceptions they may have.

Management
1. In addition to the rubber band book, students will need a variety of sources where they can find accurate information on frogs and toads and the differences between them. These sources can include encyclopedias, books, and the Internet. Please see the *Resources* section for some specific book recommendations and Internet links.
2. Students will need colored pencils in three different colors to code the facts in *Part One* as being true of frogs, toads, or both. It is suggested that you use two primary colors and the secondary color made by mixing those two colors. For example, blue for frogs, yellow for toads, and green for both frogs and toads.
3. In *Part Two*, you will need three different colors of construction paper for each group. The colors should correspond to the colors used for the fact coding in *Part One*. The 9" x 12" size paper works well.
4. Pictures of a typical frog and a typical toad are provided. Each picture should be mounted on its own piece of construction paper. You can also

use similar pictures that you cut from a book or magazine.

5. This activity may need to take place over several days in order to give students sufficient time to complete the necessary research.

Procedure

Part One—Research

1. Ask students what they know about the characteristics of frogs and toads. Discuss appearance, food, behavior, life cycle, etc. Probe to find out what differences they think exist between the two. Record all responses (even those which may contradict) on a sheet of chart paper.
2. Distribute the pages for the rubber band book and assist students with the assembly.
3. After students have read the book, revisit the responses they made. Determine if there are any incorrect statements that need to be changed, or information that should be added.
4. Distribute the first student page. Have students record characteristics discussed in the appropriate sections.
5. Ask if there are other questions that students still want to have answered about frogs and toads. (Hopefully the initial discussion will have raised some questions that were not addressed in the rubber band book.)
6. Have students get into groups. Give each group colored pencils and a copy of the second student sheet. Inform them that they will be doing some additional research on frogs and toads in order to classify each of the characteristics as belonging to only frogs, only toads, or both frogs and toads.
7. Make the research materials available to students and give them sufficient time to complete their research.
8. Have each group write down at least one additional frog and/or toad fact that they discovered in their research.

Part Two—Sorting

1. Distribute the third student page and the materials for *Part Two* to each group.
2. Have students assemble the sorting mat as described on the student page. Provide sufficient time for groups to sort and classify the characteristics. A sorting mat in progress should look something like this.

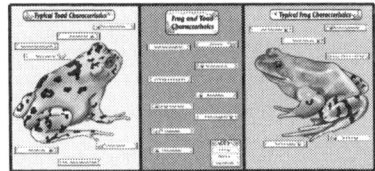

3. Have one group share the characteristics they placed on the side of the mat labeled *Typical Frog Characteristics*. See if all groups agree with the choices. Resolve any differences by consulting the research.
4. Have a second group share their *Typical Toad Characteristics* and again resolve any differences between groups. Repeat this process for *Frog and Toad Characteristics*.
5. If any group researched a characteristic that has not yet been discussed, provide time for them to share.

Connecting Learning

Part One

1. What do you know about frogs and toads?
2. How are they alike? How are they different?
3. What did you learn about frogs and toads by reading the rubber band book?
4. Are there characteristics on our list that need to be changed? Are there things that we still don't know for sure?
5. Where are some places that you can look to find the answers to your unanswered questions?

Part Two

1. What new facts did you discover about frogs and toads while doing your research?
2. Did everyone agree on where each of the characteristics belonged? If not, how did you resolve your differences?
3. What are you wondering now?

Curriculum Correlation

Berger, Melvin and Gilda. *How Do Frogs Swallow With Their Eyes? Questions and Answers About Amphibians.* Scholastic, Inc. New York. 2002.

Clarke, Barry. *Amazing Frogs & Toads.* Alfred A. Knopf. New York. 1990.

Clarke, Barry. *Eyewitness Books: Amphibian.* DK Publishing. New York. 2000.

Gibbons, Gail. *Frogs.* Holiday House. New York. 1993.

Internet Connections

Frogland
http://www.allaboutfrogs.org/weird/weird.html
This well-done site answers most conceivable questions about frogs and toads using kid-friendly language. Includes extensive links to other frog sites.

Center for Global Environmental Education
http://cgee.hamline.edu/frogs/
A site that tries to connect students, teachers, parents, and scientists in a study of frogs and toads.

Cut out this picture of a typical frog and paste it to the appropriate piece of construction paper.

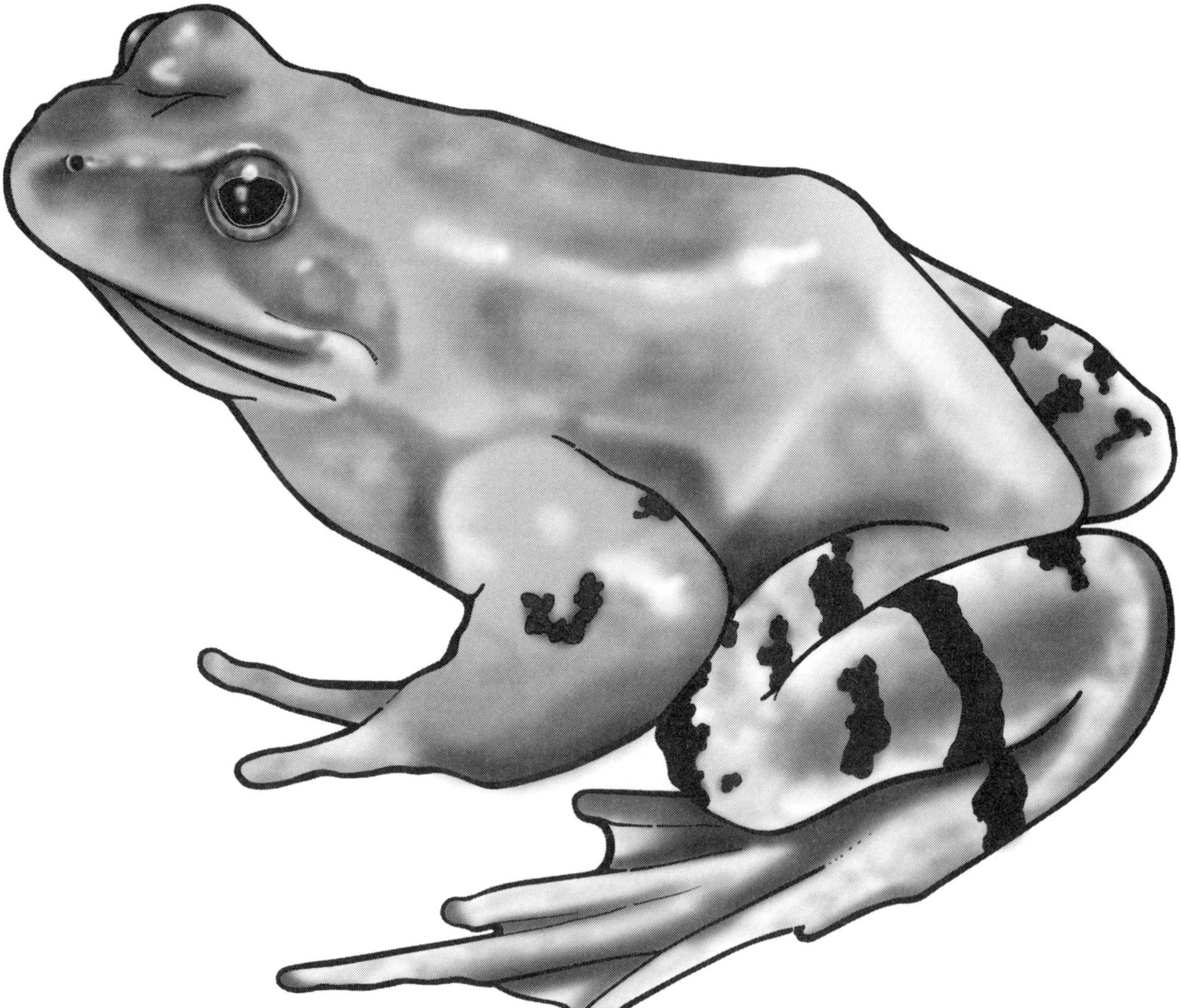

Cut out this picture of a typical toad and paste it to the appropriate piece of construction paper.

CRITTERS 121 © 2004 AIMS Education Foundation

Frog and Toad are Kin

List characteristics common to **Frogs** and **Toads**.

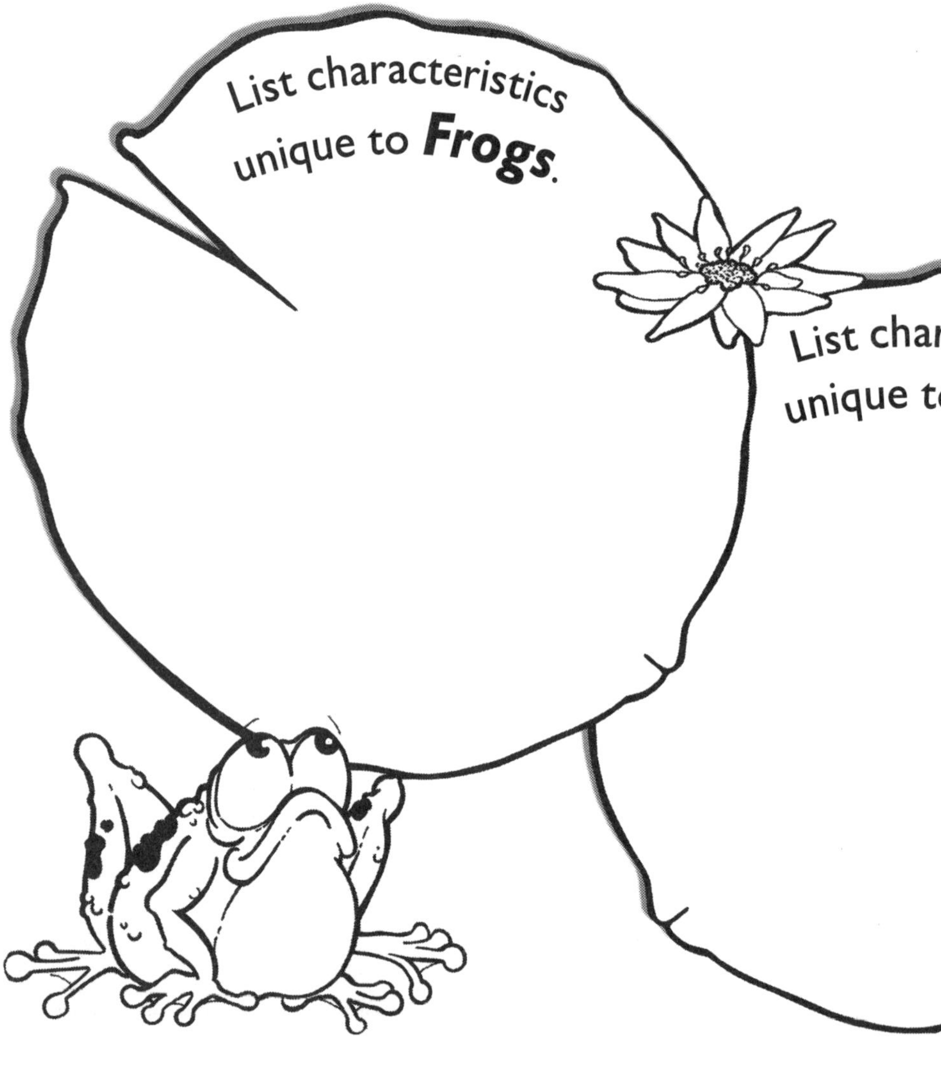

List characteristics unique to **Frogs**.

List characteristics unique to **Toads**.

Frog and Toad are Kin
Part One

Color each of the rectangles to show whether the characteristic USUALLY describes only frogs, only toads, or both frogs and toads. Use a different color for each category. (Fill in the key accordingly.) You will need to do some research to find the right answers. Use the blank spaces to record at least one new fact that you discover in your research.

- Drier, warty skin
- Drink water through skin
- Are amphibians
- Short back legs
- Feet are claw-like
- Lay eggs in long chains
- Jump up to 20 times body length
- Males call and sing to females
- Smooth, moist skin
- Have a backbone
- Long, sticky tongue
- Have poison (paratoid) glands
- Part of family *Ranidae*

- No neck and no tail
- Teeth in upper jaw
- Go through metamorphosis
- Long back legs
- Feet have webbing
- Lay eggs to reproduce
- Lay eggs in clusters
- Walk, crawl, or take short hops
- Breathe through skin
- Cold-blooded
- Eat insects and worms
- Part of family *Bufonidae*
- Carniverous

KEY
- Only Frogs
- Only Toads
- Frogs and Toads

CRITTERS 123 © 2004 AIMS Education Foundation

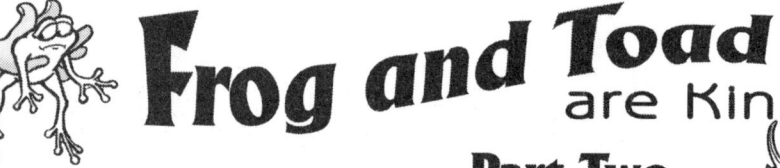

Part Two

Cut out the pictures of the frog and toad and paste each one onto a sheet of construction paper. Put the frog on the color you have been using to label frog characteristics, and the toad on the color you have been using to label toad characteristics.

Paste the toad and frog pages on either side of the remaining piece of construction paper, overlapping each page by about one inch.

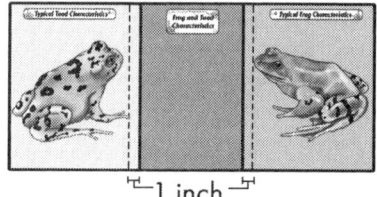

Cut out the labels and paste each one at the top of the appropriate page. Cut apart each of the frog and toad characteristics from *Part One*. Sort and classify the characteristics. Paste each one on the correct piece of construction paper. Cut out the key and paste it wherever there is room.

Labels

Under Cover

Topic
Animal coverings

Key Question
How are animal coverings different?

Learning Goal
Students will explore the various kinds of animal coverings.

Guiding Documents
Project 2061 Benchmarks
- Different plants and animals have external features that help them thrive in different kinds of places.
- Plants and animals have features that help them live in different environments.

NRC Standard
- Each plant or animal has different structures that serve different functions in growth, survival, and reproduction. For example, humans have distinct body structures for walking, holding, seeing, and talking.

Math
Counting
Graphing

Science
Life science
    animal coverings

Integrated Processes
Observing
Classifying
Comparing and contrasting

Materials
Soft material to simulate fur or hair
Feathers
Waxed paper
Net bag (see *Management 3*)
Craft sticks
Tape
Glue
Crayons
Scissors
Activity sheets

Background Information
Animals have different kinds of coverings that are adaptations to their environments. Mammals have hair or fur, birds have feathers, amphibians have smooth skins, and scales cover both fish and reptiles. These coverings serve as protection from the elements, and in some cases, other animals. They allow the animals to best survive in the environments in which they live.

Management
1. Fur can be simulated by small patches of soft, fur-like material, or by gluing hair collected from pets or a dog groomer onto small squares of paper.
2. Feathers can be purchased at a craft store or obtained from a feather pillow.
3. Scales can be simulated by cutting small patches from net produce bags (like the ones used for onions or oranges).
4. Squares of waxed paper can be used to simulate smooth skin.
5. In sorting and graphing the small animal pictures, there are too many animals with fur to fit on the graph. This provides an opportunity for students to problem solve, but depending on the age of your students, you may wish to eliminate two furry animals before distributing the pictures.
6. This activity can be done in one or two sessions.

Procedure
1. Discuss animal coverings. Then hand out the two activity sheets with the four large animal drawings on them and have the students cut them out.
2. Have students color in the animals and glue a patch of the appropriate skin covering on each one.
3. Instruct students to tape a craft stick to the back of each picture, making an animal puppet.
4. Tell students that they will use the puppets to answer questions about animal coverings. For example, they would hold up the snake puppet if you showed the class a picture of a fish and asked what kind of skin covering a fish has.
5. After the students have a good grasp of what kind of animals have the various coverings, distribute the activity sheets with the covering graph and the small animal pictures.
6. Inform students that they are to glue a sample of each covering in the appropriate box on the graph.
7. Once this has been accomplished, have students cut out the small animal pictures and glue them in the appropriate squares on the graph.

CRITTERS

Connecting Learning
1. Why do animals need coverings? [to protect them from weather, to protect them from other animals, etc.]
2. What are some different animal coverings? [fur, scales, feathers, smooth skin, etc.]
3. How does _____ (covering) meet the needs of _____ (animal)?
4. What are some other kinds of coverings that animals have? [hard shell of a snail, thick skin of an elephant, etc.]
5. What are you wondering now?

Extensions
1. Make "feely" bags that simulate the different textures of animal coverings.
2. Children can bring in toy animals to sort according to covering types or to make a class "zoo."

Curriculum Correlation
Literature
Goodman, Susan. *Claws, Coats, and Camouflage.* Millbrook Press. Brookfield, CT. 2001.

Language Arts
Brainstorm furry things, then write and illustrate furry stories or poems.

CRITTERS 128 © 2004 AIMS Education Foundation

Under Cover

Fur	Feathers	Scales	Smooth Skin

Under Cover

Wonderful Webbed Feet

Topic
Adaptations: webbed feet

Key Questions
How does the webbing in frogs' feet help them swim?

Learning Goals
Students will:
1. learn about the variety in the structure of the feet of different types of frogs,
2. make paddle boats that simulate two different kinds of frog feet,
3. use webbed and non-webbed hands to compare motion through water, and
4. generalize what kind of foot is best for frogs that live in the water.

Guiding Documents
Project 2061 Benchmarks
- *Some animals and plants are alike in the way they look and in the things they do, and others are very different from one another.*
- *Plants and animals have features that help them live in different environments.*
- *Different plants and animals have external features that help them thrive in different kinds of places.*

NRC Standard
- *Each plant or animal has different structures that serve different functions in growth, survival, and reproduction. For example, humans have distinct body structures for walking, holding, seeing, and talking.*

NCTM Standard 2000*
- *Use tools to measure*

Math
Measurement
 linear

Science
Life science
 adaptations
 frog feet

Integrated Processes
Observing
Predicting
Collecting and recording data
Comparing and contrasting
Relating
Applying
Generalizing

Materials
For each pair of students:
 2 or 3 Styrofoam meat trays (see *Management 1*)
 2 or 3 rubber bands (see *Management 3*)
 2 large paper clips (see *Management 5*)
 measuring tools (see *Management 6*)
 one pair of clear plastic gloves
  (see *Management 9*)
 copies of frog foot picture cards
 scissors

For the class:
 several tubs of water (see *Management 4*)
 clear plastic wrap (see *Management 8*)
 clear packaging tape

Optional:
 non-fiction books and pictures of frogs showing their feet
 Internet access (see *Internet Connections*)

Background Information
 An adaptation is a characteristic of an animal or plant that enables it to live successfully in a particular place. Adaptations may be physical, or they may be part of an animal's behavior. The animals and plants that are best adapted to their environments are most likely to survive and reproduce, passing along their adaptive characteristics to their offspring.
 Frogs all have four toes on their front feet and five toes on their back feet. Beyond that, the design of the feet varies greatly, providing excellent examples of adaptations. A close look at the legs and feet of a particular type of frog will tell a great deal about how it lives its life, its behavior, and the habitat in which it lives. Frogs that live mostly on land usually have shorter legs, which are better suited for walking and climbing. Some of these frogs have broad feet

with short, stubby toes, ideal for burrowing into the ground. Tree-climbing frogs have large, round, sticky pads on their toes that work like suction cups to help them cling to branches, leaves, and other often slippery surfaces. Water-dwelling frogs have comparatively long, strong legs with webbed feet that help them swim better and faster.

Webbing of feet is an adaptation shared by ducks, geese, swans, otters, salamanders, frogs, and certain other animals that spend a significant part of their lives in water. In this activity, students will experience the advantage of having webbed feet as an animal moves through water. In *Part One*, they will make paddle boats with two different kinds of paddles representing webbed and non-webbed feet. In *Part Two*, they will experience the effects of webbing as they move their own gloved hands, one of which is webbed, through a container of water.

Management

1. You can get Styrofoam meat trays at a grocery store or use Styrofoam take-out boxes. Styrofoam plates should be used only if they are sturdy; many are too flimsy and will buckle under the pressure of the rubber band.
2. Depending upon the cutting skills of your students, you may need to cut the paddles ahead of time or assist the children in doing so themselves.
3. Rubber bands should be medium in size and fairly thin. You may need to test several sizes to find the ones that work best.
4. If available, a small wading pool works very well. Otherwise, use individual tubs at least four inches deep and 24 inches long (longer is better).
5. Use a large paper clip on the front of the boat to keep it from flipping over.
6. Those who are comfortable with customary units may use rulers to measure the distance traveled, those who are not can use string.
7. Several non-fiction frog books and/or pictures should be available for looking closely at the structure of frogs, especially their feet. See also the websites listed.
8. Before doing *Part Two*, cut the clear plastic wrap into 6-inch by 3-inch strips—one for each student.
9. Inexpensive disposable clear plastic gloves used for handling food, etc., may be available from the school cafeteria or purchased at discount stores.

Procedure

Part One
1. Use pictures, books, or images from the Internet to show some of the variety in the structure of the feet of different types of frogs. Ask the students to offer suggestions for the purposes of the different feet they see. Call the students' attention to the differences between front and back feet as well. (See *Background Information*.)
2. Tell the students that they will be making boats that are propelled in a way similar to how the frogs' feet propel them through the water. The boats will have two different paddles representing two of the different types of frog feet.
3. Have students get into pairs and distribute the materials. Go over the construction instructions as a class. Assist students as necessary during boat construction.
4. Demonstrate the method of sailing the boat with the webbed paddle:

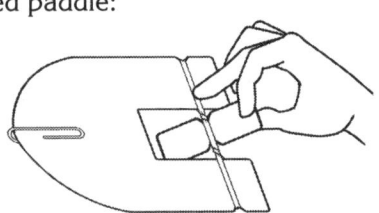

   (a) Be sure the paddle is centered in the rubber band.
   (b) Rotate the paddle from the top towards yourself.
   (c) Use five twists of the paddle for consistency.
   (d) Use one hand to launch the boat. It helps to push down very slightly so that the front of the boat is just a little bit above the surface of the water.
5. Have the students launch the web-paddled boat and observe. Discuss the results, including distance traveled, speed, course, and anything else they notice. Allow time for the students to troubleshoot and fix problems. If the boats do not go in a relatively straight line, the paddle may be hitting the side of the slot. If the boat goes backwards, the paddle was wound in the wrong direction. If it flips, another paper clip may need to be added to the front, or the students may be pushing down too hard on the back of the boat. If necessary, launch the boats again until the students are successful.
6. Ask the students to predict what might happen using the non-webbed paddle. Repeat *Procedure 5* using that paddle. Compare the results, discuss, and record on the journal page.
7. Launch both boats side by side, comparing the speed and distance traveled. Encourage the students to do several test launches until their results are consistent.
8. Bring the class together around one tub. Have one team launch two boats side by side. Ask the students to reflect upon which boat(s) traveled faster and farther, and what difference was made by each type of paddle.
9. Have students look at the pictures of the frogs' feet and ask the students which paddle looks most like which kind of foot. Talk about what the advantages of having webbed feet might be for a frog that lives primarily in the water.

Part Two
1. Remind the students of what they observed in *Part One*. Explain that in this part of the activity, they will experience firsthand what it feels like to have webbed feet (hands).
2. Distribute two pairs of gloves to each team. Have one student in each team put on a pair of gloves. Have him or her spread the fingers on one hand as far apart as possible. Direct the other student to cover the four fingers with plastic wrap and tape it securely, forming webbing.

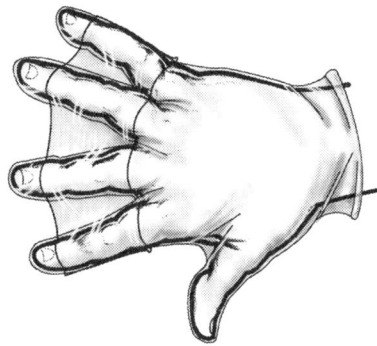

3. Now tell the student wearing the gloves to drag both hands through the tub of water with fingers on both hands spread wide apart. Ask him or her to compare what he/she feels with each hand, webbed and non-webbed, while the other student observes and describes the motion of the water.
4. Instruct the partners to switch places and repeat the procedure, then record their observations on the journal page.
5. Bring the class together again around a tub and choose one team to demonstrate. Ask the students to share their observations. Discuss the advantage of having webbed feet for frogs and certain other water-dwelling animals. Encourage students who have used swim fins to relate that experience with what they have done in this activity.

Connecting Learning
1. Looking at the pictures, how are the frogs' feet different? How would each of these types of feet help the frog to live in its habitat?
2. How are the boat paddles similar to different kinds of frog feet?
3. Which kind of paddle moved the boat farthest and the fastest? Why? Did the other teams have the same results you did? Why or why not?
4. Describe how your webbed hand felt when it moved through the water. ...your non-webbed hand. What difference did the webbing make? What did it do to the water?
5. How did the boats' paddles help them move? How do a frog's feet help it swim?
6. If you were to design a foot for a water-dwelling animal, what would it look like? Explain your thinking.
7. Which would make a better paddle for a boat—a fork or a spoon? Explain.
8. What are you wondering now?

Extensions
1. Find a toy rubber ball covered with suction cups. Use it to illustrate the foot adaptations of climbing frogs.
2. Research other animals with webbed feet.
3. Encourage students to improve upon the design of the boat paddle.

Internet Connections
Excellent information about adaptations in frogs:
http://www.exploratorium.edu/frogs/mainstory/frogstory2.html

Frog Links:
http://www.exploratorium.edu/frogs/links.html

Curriculum Correlation
Clarke, Barry. *Eyewitness Books: Amphibian*. DK Publishing. New York. 2000.

Frogs (Face-to-Face). Scholastic, Inc. New York. 2001.

Gibbons, Gail. *Frogs*. Holiday House. New York. 1994.

Lovett, Sarah. *Extremely Weird Frogs*. Avalon Travel Publishing, John Muir. New York. 1996.

Pallotta, Jerry. *The Frog Alphabet Book*. Charlesbridge Publishing. Boston. 1990.

Parsons, Harry. *The Nature of Frogs: Amphibians with Attitude*. GreyStone Books. Berkley, CA. 2000.

Home Link
Encourage students to take their boats home to demonstrate and continue using.

* Reprinted with permission from *Principles and Standards for School Mathematics*, 2000 by the National Council of Teachers of Mathematics. All rights reserved.

Wonderful Webbed Feet

Making the Boat Pieces
- Carefully cut out each pattern.
- Trace the patterns onto the Styrofoam plates or trays.
- Trace two boat patterns.
- Cut out the two boats and the two paddles.

Assembling the Boats
- Put a large paper clip on the nose of each boat.
- Put a rubber band around each boat so that it fits in the notches.
- Slide one of the paddles into the cut out section of one boat. Put it between the rubber bands so that the notches line up.
- Twist the paddle towards you five times.
- Carefully put the boat into the water and let go of the paddle.
- Repeat with the other paddle and the other boat.

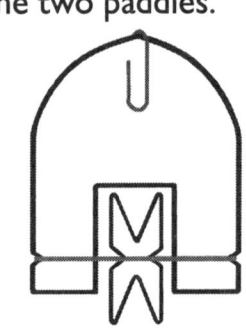

Paddle Patterns

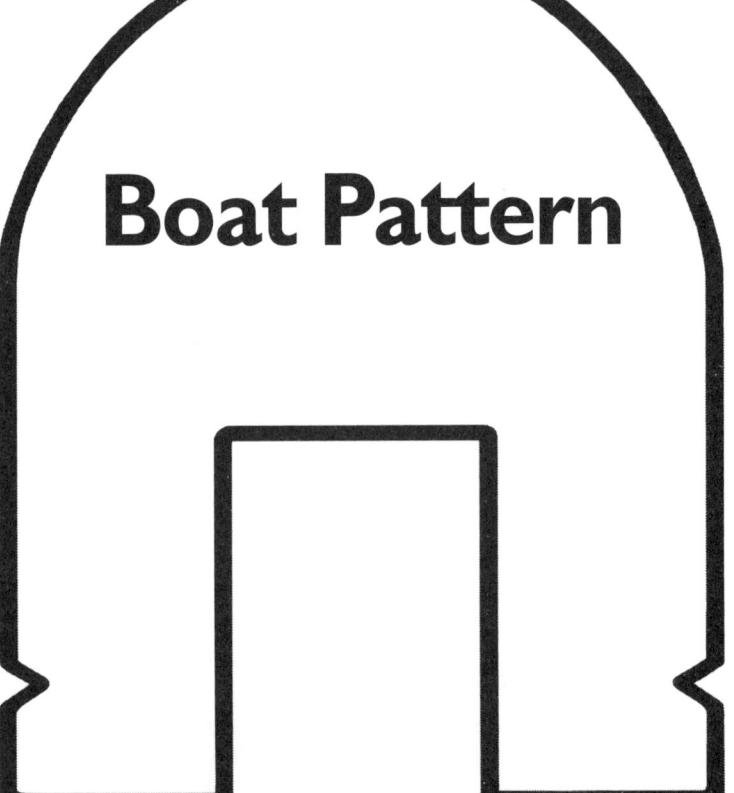

Boat Pattern

CRITTERS © 2004 AIMS Education Foundation

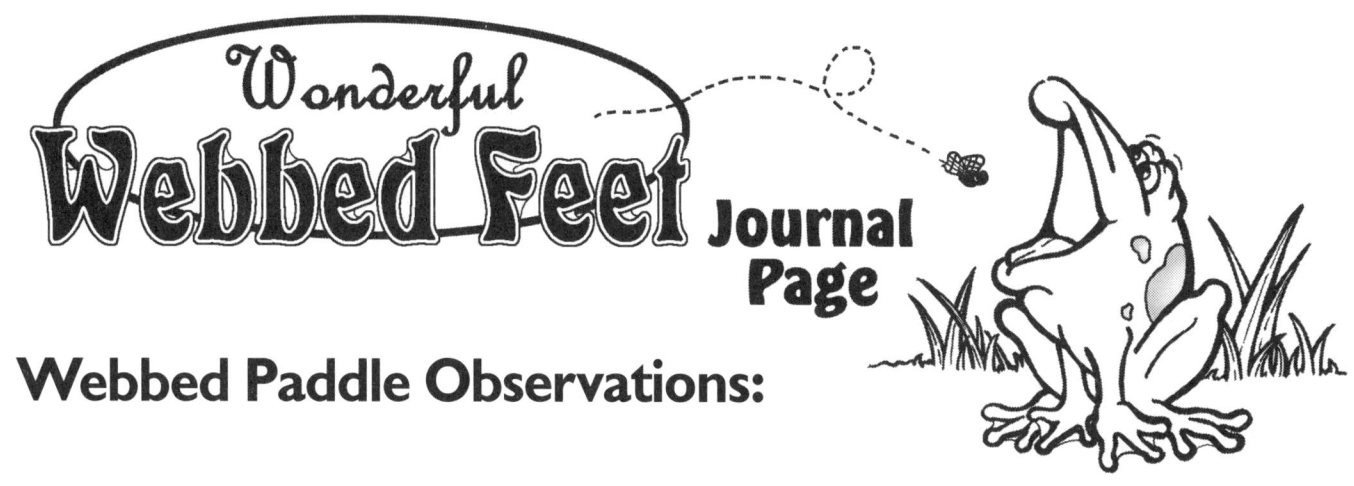

Wonderful Webbed Feet Journal Page

Webbed Paddle Observations:

Non-webbed Paddle Observations:

Glove Obserations:

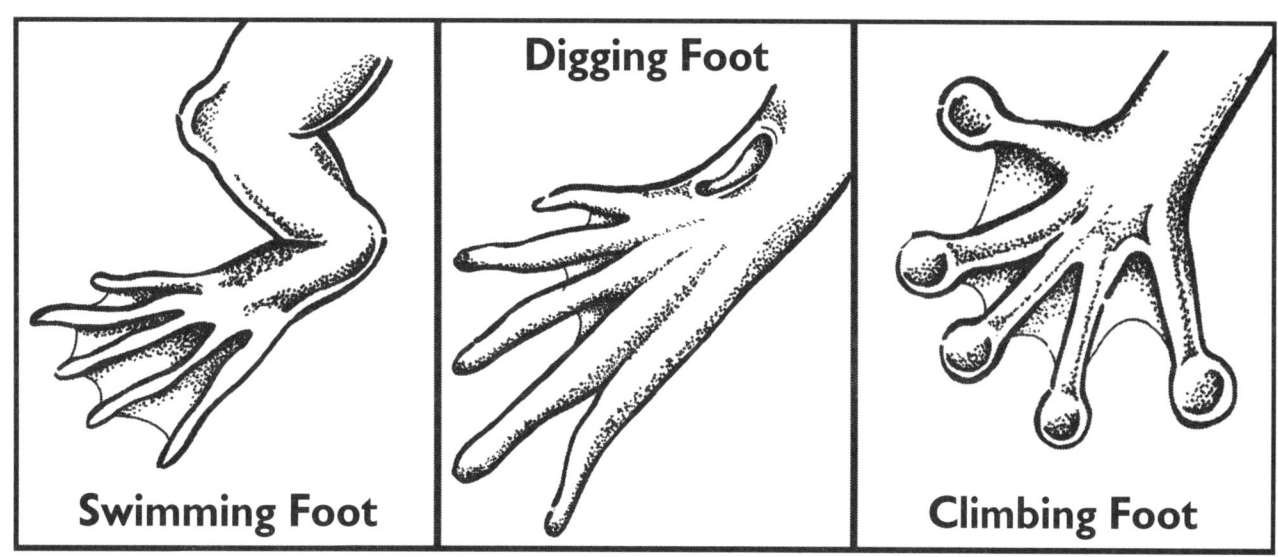

Swimming Foot | Digging Foot | Climbing Foot

Critters Hide 'n' Seek

Topic
Camouflage

Key Question
How does camouflage affect a critter's ability to survive?

Learning Goals
Students will:
1. make a critter, and
2. see the effects of camouflage on animal visibility.

Guiding Documents
Project 2061 Benchmark
- *For any particular environment, some kinds of plants and animals survive well, some survive less well, and some cannot survive at all.*

NRC Standard
- *An organism's patterns of behavior are related to the nature of that organism's environment, including the kinds and numbers of other organisms present, the availability of food and resources, and the physical characteristics of the environment. When the environment changes, some plants and animals survive and reproduce, and others die or move to new locations.*

Math
Counting
Fractions

Science
Life science
   animals
      camouflage
      predator/prey relationships

Integrated Processes
Observing
Predicting

Materials
Camouflage materials
  (see *Management 3*)
Critter-building materials
  (see *Management 4*)

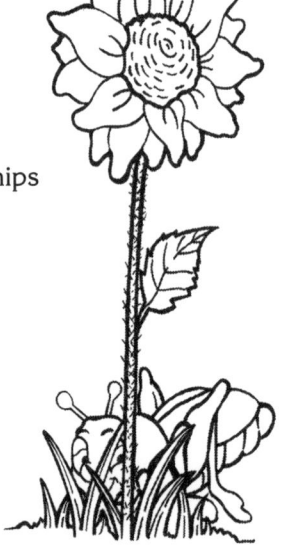

Background Information
Animals use camouflage to protect themselves from predators. Color and body shape are two important variables that enable an animal to blend into its environment. Animals have adapted to the predominant colors in their surroundings. An animal's body shape and size affect its ability to blend into the environment. The patterns of colors on the body covering also affect the animal's chances of being seen.

Management
1. Be sure to choose an area beforehand where students can hide their critters. This activity woks best if the chosen area is not too large and has definite boundaries.
2. Emphasize that students may not bury the critters.
3. Use tempera paint, markers, or other materials for camouflage.
4. For the critter bodies use paper bags, marshmallows, cardboard egg carton parts, clay, or wads of paper glued or taped together. Toothpicks or sticks can be used for arms and legs.

Procedure
1. Discuss the *Key Question* and activity sheet. Note that the activity sheet is in outline form and is intended to integrate science and language arts.
2. Send students outside to select and observe an environment for their critters within the area chosen for this activity.
3. Have students record environment information on the activity sheet.
4. Keeping environmental information in mind, have students create and camouflage a critter using the materials available.
5. Have them record the vital statistics and critter facts on the activity sheet.
6. Have students draw pictures of their critters.
7. Divide the class into two groups and send one group outside to hide their critters in the selected environment. Remind students that they may not bury their critters.
8. Have students make predictions on the second activity sheet.

9. Send the second group of "predator" students outside to try and find the first groups' critters. Allow the predators to make one or two sweeps of the area. Set a time limit of a few minutes.
10. Reverse the groups so that the second group has the opportunity to hide their critters and the first group has the opportunity to be predators.
11. Have students complete the data on the activity sheets, making fractions of the number of critters found and not found to the number hidden.
12. Collect all the critters and divide them into two groups—found and not found. Make a real graph of the data.

Connecting Learning
1. Were you able to locate all of the critters? Why or why not?
2. Look at the critters in the found and not found groups. What do the critters that were found have in common?
3. What do the critters that were not found have in common?
4. If you were to do this again, what would you change about your critter?
5. Why would a critter's coloring be important to its survival?
6. Can you think of an animal that has protective coloring? Explain.
7. What are you wondering now?

Extensions
1. Emphasize the use of shapes for camouflage. Use green and brown construction paper.
2. Select a different environment or season and adapt your critter.

Curriculum Correlation
Arnosky, Jim. *I See Animals Hiding.* Scholastic, Inc. New York. 2000.

Goodman, Susan. *Claws, Coats, and Camouflage.* Millbrook Press. Brookfield, CT. 2001.

Powzyk, Joyce. *Animal Camouflage: A Closer Look.* Bradbury Press. New York. 1990.

Critters
Hide 'n' Seek

I. Enviroment
   A. Colors
      1. _____
      2. _____
   B. Shapes
      1. _____
      2. _____

II. Vital Statistics
   A. Name _____
   B. Height _____
   C. Length _____
   D. Mass _____

III. Critter Facts
   A. Predators _____
   B. Defense _____
   C. Camouflage _____
   D. Food _____

Draw your critter here.

My Guess

How many critters will be found? _____

How many critters will not be found? _____

Data

Number of critters hidden? _____

Number of critters found? _____

Number of critters not found? _____

Fractions

Write a fraction to describe the number of critters found and not found.

$$\frac{B}{A} = \frac{\#\ found}{\#\ hidden} = \underline{\qquad\qquad}$$

$$\frac{C}{A} = \frac{\#\ not\ found}{\#\ not\ hidden} = \underline{\qquad\qquad}$$

Gone Fishing

Topic
Camouflage

Key Question
How does the color of an animal affect its population?

Learning Goal
Students will use paper fish cutouts to see the effect of camouflage on prey populations.

Guiding Documents
Project 2061 Benchmarks
- Different plants and animals have external features that help them thrive in different kinds of places.
- Individuals of the same kind differ in their characteristics, and sometimes the differences give individuals an advantage in surviving and reproducing.
- For any particular environment, some kinds of plants and animals survive well, some survive less well, and some cannot survive at all.
- Plants and animals have features that help them live in different environments.
- Use fractions and decimals, translating when necessary between decimals and commonly encountered fractions—halves, thirds, fourths, fifths, tenths, and hundredths (but not sixths, sevenths, etc.)

NRC Standard
- Each plant or animal has different structures that serve different functions in growth, survival, and reproduction. For example, humans have distinct body structures for walking, holding, seeing, and talking.

*NCTM Standards 2000**
- Collect data using observations, surveys, and experiments
- Represent data using tables and graphs such as line plots, bar graphs, and line graphs

Math
Sequencing
Fractions
Computation
Equalities and inequalities
Graphing

Science
Life science
   camouflage

Integrated Processes
Observing
Predicting
Comparing and contrasting
Collecting and recording data
Drawing conclusions

Materials
Per group of four:
   blue, red, black, and white construction paper
   construction paper fish (see *Management 2*)
   watch with second hand
   activity sheets

Background Information
   For many creatures, color is an important means of defense. The blending of an animal into its environment is called camouflage. Camouflage is one way an animal adapts to its environment. The snowshoe hare is white during the winter so that it can blend in with the snow. When the snow melts, the color of the hare changes to brown so that it blends into the surroundings during the summer months too. Some animals, like the walking stick and tomato worm, blend so well into their surroundings that they are undetectable when motionless.

Management
1. The fish need to be cut out prior to doing this activity. Students can use the pattern at the top of the second activity sheet or cut out their own fish pattern.
2. The fish need to be the same size and shape and cut from black, red, white and blue construction paper. Each group will need 12 of each color for the activity.
3. Make a few extra fish in case some get torn during the activity.
4. Small rectangles of paper (2" x 1") can be used instead of the fish shapes.
5. This activity should be done before doing *Missing Moths*.

CRITTERS   140   © 2004 AIMS Education Foundation

Procedure
1. Distribute the activity sheets.
2. Divide the class into groups of four and make sure each group has the cut-out fish and two sheets of blue paper.
3. Have groups spread the blue paper on a table to act as the fishing pond. Instruct the first "fisherman" to turn his/her back to the pond while the other group members spread the 48 fish evenly over the blue paper.
4. Explain the rules of the activity: Each fishing period will last exactly 10 seconds. The fisherman may only use one hand to pick up fish from the paper. The collected fish may be put in a cup or held in the other hand. The fisherman must attempt to get as many fish as possible in 10 seconds.
5. Have groups time the first fisherman. At the end of 10 seconds, instruct them to count and record the number of fish of each color that were caught.
6. Repeat this procedure until all four people have had a chance to fish. Make sure that all the fish are replaced before starting each new fishing period.
7. Have the groups share their data and make a data chart on the board.
8. Allow time for students to complete the final two student pages. Discuss the results.

Connecting Learning
1. Which color of fish was caught the most often? Why do you think this is?
2. Which color of fish was caught the least often? Why do you think this is?
3. How did your group data compare to the class data?
4. How does this activity relate to animals living in their natural environment?
5. What would happen to snowshoe hares if there was no snow one winter and the ground stayed brown?
6. What are you wondering now?

Extensions
1. Use other colors of fish and backgrounds and repeat the activity.
2. Repeat the activity with the same colors, but different sizes of fish, to see if size makes a difference.

Curriculum Correlation
Literature
Arnosky, Jim. *I See Animals Hiding*. Scholastic, Inc. New York. 2000.

Goodman, Susan. *Claws, Coats, and Camouflage*. Millbrook Press. Brookfield, CT. 2001.

Powzyk, Joyce. *Animal Camouflage: A Closer Look*. Bradbury Press. New York. 1990.

Language Arts
Write a story about a brown mouse that lived in a white environment.

Art
Do a camouflage art lesson in which the students draw pictures with objects hidden in them.

Research
Find other animals that use camouflage.

* Reprinted with permission from *Principles and Standards for School Mathematics*, 2000 by the National Council of Teachers of Mathematics. All rights reserved.

Gone Fishing

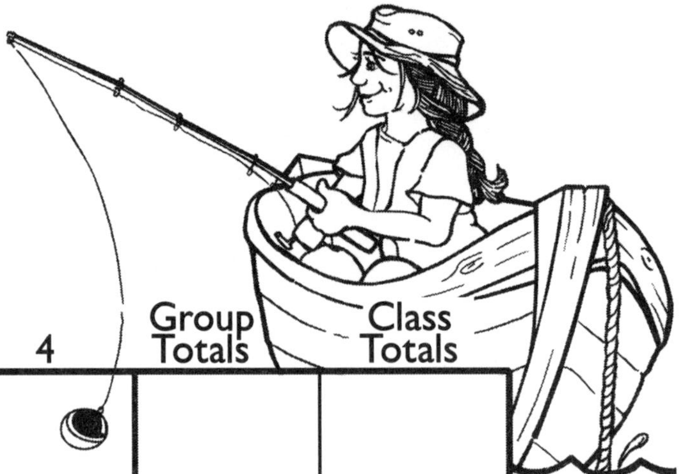

Counting our catch

	Group Members 1	2	3	4	Group Totals	Class Totals
White						
Red						
Blue						
Black						

Looking at our catch

1. Which fish was caught the most? _____
   ...the least? _____

2. Label the fish in order from the most caught to the least caught.

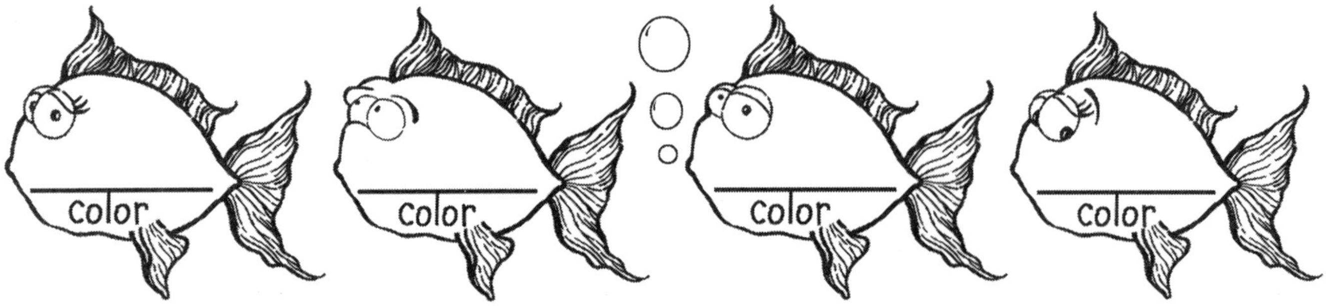

color _____   color _____   color _____   color _____

Number caught   Number caught   Number caught   Number caught

CRITTERS

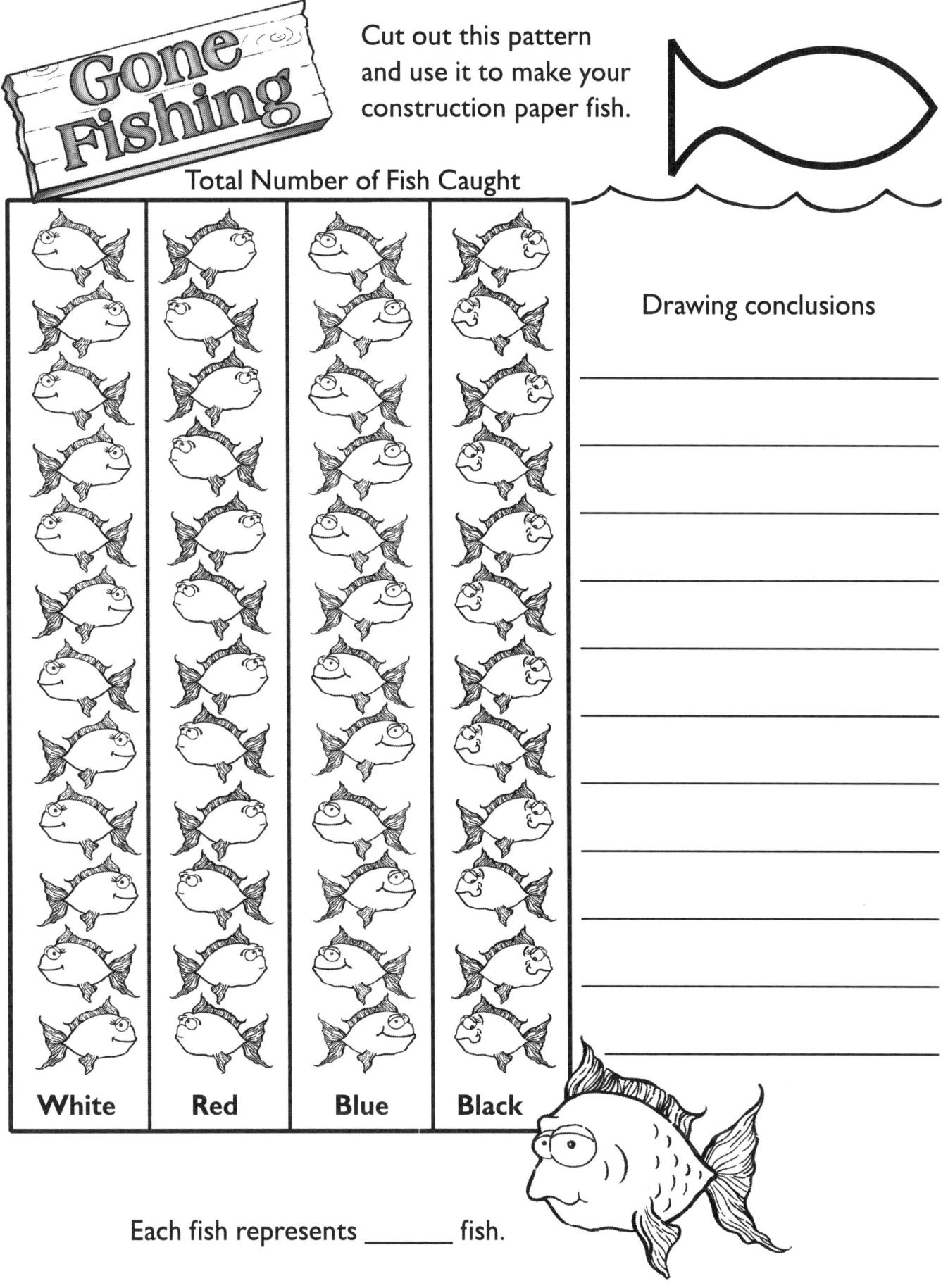

Gone Fishing

	Fisherman's Total	Total # of Fish	Fraction Caught	Fraction Not Caught
White		12		
Red		12		
Blue		12		
Black		12		

Use >, =, < symbols to record your results.

The fraction of red fish caught is ◯ the fraction of white caught.

The fraction of blue fish caught is ◯ the fraction of black caught.

The fraction of black fish not caught is ◯ the fraction of red fish not caught.

Now fill in the blanks.

The fraction of _____ caught is > the fraction of _____ caught.

The fraction of _____ not caught is < the fraction of _____ not caught.

Write your own sentence about the number of fish caught.

Missing Moths

Topics
Camouflage
Adaptations

Key Questions
1. How does camouflage affect an animal's ability to be seen?
2. How does a moth's color affect its ability to be located on a coordinate map?

Learning Goals
The students will:
1. observe an environment with a variety of moths to see the effects of camouflage on animal visibility,
2. design their own camouflaged moths,
3. apply this knowledge to the sphinx moth of England and its rapid adaptation during the industrial revolution,
4. try to locate moths camouflaged on a coordinate grid, and
5. map the moths' locations.

Guiding Documents
Project 2061 Benchmarks
- For any particular environment, some kinds of plants and animals survive well, some survive less well, and some cannot survive at all.
- Different plants and animals have external features that help them thrive in different kinds of places.
- Plants and animals have features that help them live in different environments.
- Changes in an organism's habitat are sometimes beneficial to it and sometimes harmful.
- Individuals of the same kind differ in their characteristics, and sometimes the differences give individuals an advantage in surviving and reproducing.

NRC Standards
- Biological evolution accounts for the diversity of species developed through gradual processes over many generations. Species acquire many of their unique characteristics through biological adaptation, which involves the selection of naturally occurring variations in populations. Biological adaptations include changes in structures, behaviors, or physiology that enhance survival and reproductive success in a particular environment.
- Each plant or animal has different structures that serve different functions in growth, survival, and reproduction. For example, humans have distinct body structures for walking, holding, seeing, and talking.

NCTM Standards 2000*
- Collect data using observations, surveys, and experiments
- Represent data using tables and graphs such as line plots, bar graphs, and line graphs
- Make and use coordinate systems to specify locations and to describe paths
- Describe location and movement using common language and geometric vocabulary

Math
Counting
Estimation
Coordinate mapping
Graphing
Computation
   fractions
   decimals
   percents

Science
Life science
   adaptations
      protective coloration

Integrated Processes
Observing
Comparing and contrasting
Collecting and recording data
Interpreting data
Drawing conclusions
Applying

Materials
For the class:
   newspaper want ads
   construction paper (brown, green, and white)
   butcher paper
   scissors
   tape
   meter stick
   black felt-tip marker

For each student:
  colored pens or pencils
  activity sheets

Background Information

An animal's ability to blend into an environment is called camouflage. Camouflage can be used for defensive or offensive purposes. A rabbit uses camouflage to hide from predators. A mountain lion uses it to hide until a prey is close enough to attack.

Both color and shape can camouflage animals. A walking stick is an example of this. Its shape and color make it appear to be part of a tree branch.

One of the most dramatic cases of an animal's response to a changing environment is the sphinx moth of England. Prior to 1850, the sphinx moths that lived near Manchester were light colored, making them blend into the light-colored bark of the surrounding trees. By 1894, 95% of the sphinx moths were dark colored. This change occurred because of the environmental effects of the industrial revolution. The local industries were burning large amounts of fuel that produced a new phenomenon, air pollution. The vegetation became coated with this pollution and turned darker in color. The light-colored moths became highly visible on the darkened trees and were easy targets for their predators. In 1848 the first black moth was captured, and, in 47 years, successive populations of moths had changed their color to adapt to the darker environment. The darker moths are presently more populous than the lighter ones. However, due to ecological efforts and the use of cleaner fuels, the light-colored moth is beginning to make a comeback. This case is unusual because coloration changes generally occur over a much longer period of time. This example clearly illustrates the impact humans have made on the environment.

Time Line

	Industrial revolution		Environment improves
1800	1850	1900	1950
All moths are light	First dark moth found	Most moths (95%) have become dark	Light moths increasing

Management
1. Before doing the lesson, prepare the "moth environment." To do this, use a double page of the want ads. Be sure the pages are covered with small print. Use the moth pattern to cut out brown, green, white, and newsprint (cut from the want ads) moths. (If you have access to an Ellison machine, you may find it easier to use a die cut to make the moths.) Randomly glue the moths onto the double page of the wants ads. The numbers of each type of moth may vary.
2. You may want to cut the newsprint moths out of another identical page of want ads and glue them in the exact position from which they were cut out, making them very well camouflaged.
3. Laminate the sheets if desired.
4. Copy one moth pattern on white paper for each student.
5. For *Part Two*, you will need to make a second moth environment or use the environment from *Part One*. Divide this environment into 16 coordinate sections as shown on the activity sheet for *Part Two*. Use a marker and a meter stick to draw the grid lines.

Procedure
Part One
1. Before the students arrive, place the newspaper with the moths glued to it on a wall, bulletin board, or chalkboard in the front of the classroom and cover it with a sheet of butcher paper.
2. To begin the lesson, tell the class that you have a page of paper moths under the butcher paper. Their task will be to look at the paper for 15 seconds and estimate the total number of moths and the number of different types (colors) of moths.
3. Distribute the student activity sheet for *Part One*.
4. Remove the butcher paper and allow students to observe the paper for 15 seconds. Re-cover it with the butcher paper.
5. Have the students complete the first section by recording their estimates of the number of types of moths and the total number of moths they saw. Discuss their estimations.
6. Uncover the paper and count the actual number of types and the total number of moths. Record this data in section two of the activity sheet.
7. Discuss how the predictions and results compared. Because of camouflage, many students may not have seen the newsprint moths. This should be related to how animals depend on protective coloration to help them survive.
8. Complete section three by having the students count and record the number of each type of moth.
9. Complete the bar graph by asking the students to raise their hands when you call out the color of the moth that was easiest for them to see. Record this data on the board and have the students color in their graphs accordingly.
10. Tell them that they are going to choose a spot in the room to place (tape) their moth. Distribute one moth pattern to each student.
11. Invite each student to select a place and color his or her moth so that it will blend into that location and be camouflaged. Encourage students to

consider both the colors and the patterns in the areas they choose.
12. Have the students leave the classroom and re-enter one at a time and tape their moths to their chosen locations. Note: Moths must be in plain sight and not be placed under anything.
13. Have all students return and look around to see how many moths they can see.
14. Tell students the story of the sphinx moth (see *Background Information*).

Part Two
1. Display the second moth environment at the front of the classroom.
2. Have the students sit so that they are at least two or three meters from the moth environment.
3. Hand out the activity sheets for *Part Two* and discuss. Students are to decide on a key for describing the moths, i.e., Brown = B, Classified = C, etc.
4. Using their keys, have students map the locations of the moths in the moth environment on the activity sheet grid. Do not allow the students to go near the sheet for a closer look. Tell them to mark all of the moths they see on the grid.
5. Point out all of the moths to the class and have them record the actual number of each kind by drawing them into the grid and circling them.
6. Have students write a conclusion about the effectiveness of camouflage.
7. Discuss these conclusions.

Connecting Learning
Part One
1. Which moths were the most camouflaged? ...the least camouflaged?
2. What would happen if the background had been red? ...black? ...white?
3. How does this example relate to animals in the forest?
4. Why are different moths more easily seen by some of our class?
5. Which of your classmates' moths were you unable to spot? Why? Were others able to spot your moth? Why or why not?
6. Do you think that the sphinx moth changed rapidly? Explain your thinking. [The sphinx moth took between 50 to 100 years to become mostly dark. While on a large scale, this seems rapid, for a moth, this represents many, many generations.]
7. What do you think would have happened to the sphinx moth had it not changed color during the industrial revolution?
8. What are you wondering now?

Part Two
1. How did the moth's color affect its ability to be seen on the grid?
2. Which color was the easiest to locate?
3. Which color was the most difficult to locate?
4. Were there factors other than camouflage that affected your ability to locate each moth?

Extensions
1. Use different backgrounds (white, green) for moth environments. Compare results with those using newsprint.
2. Use different colors of moths, or make the moths from another material, such as felt.
3. Have students make camouflage books using old wallpaper patterns. Expired wallpaper books can be obtained from home improvement and home decorating stores.
4. Do this activity on the grass in a marked off area using various shades of green moths.

Curriculum Correlation
Literature
Arnosky, Jim. *I See Animals Hiding*. Scholastic, Inc. New York. 1995.

Goodman, Susan. *Claws, Coats, and Camouflage*. Millbrook Press. Brookfield, CT. 2001.

Powzyk, Joyce. *Animal Camouflage: A Closer Look*. Bradbury Press. New York. 1990.

Whalley, Paul. *Butterfly & Moth (Eyewitness Books)*. DK Publishing, Inc. New York. 2000.

Math
Make a coordinate map of your classroom showing where each desk is located.

Computer
Move the turtle in LOGO by giving it coordinates and using the set position commands.

Social Sciences
Use a United States map and identify the coordinates of each state's capital.

* Reprinted with permission from *Principles and Standards for School Mathematics,* 2000 by the National Council of Teachers of Mathematics. All rights reserved.

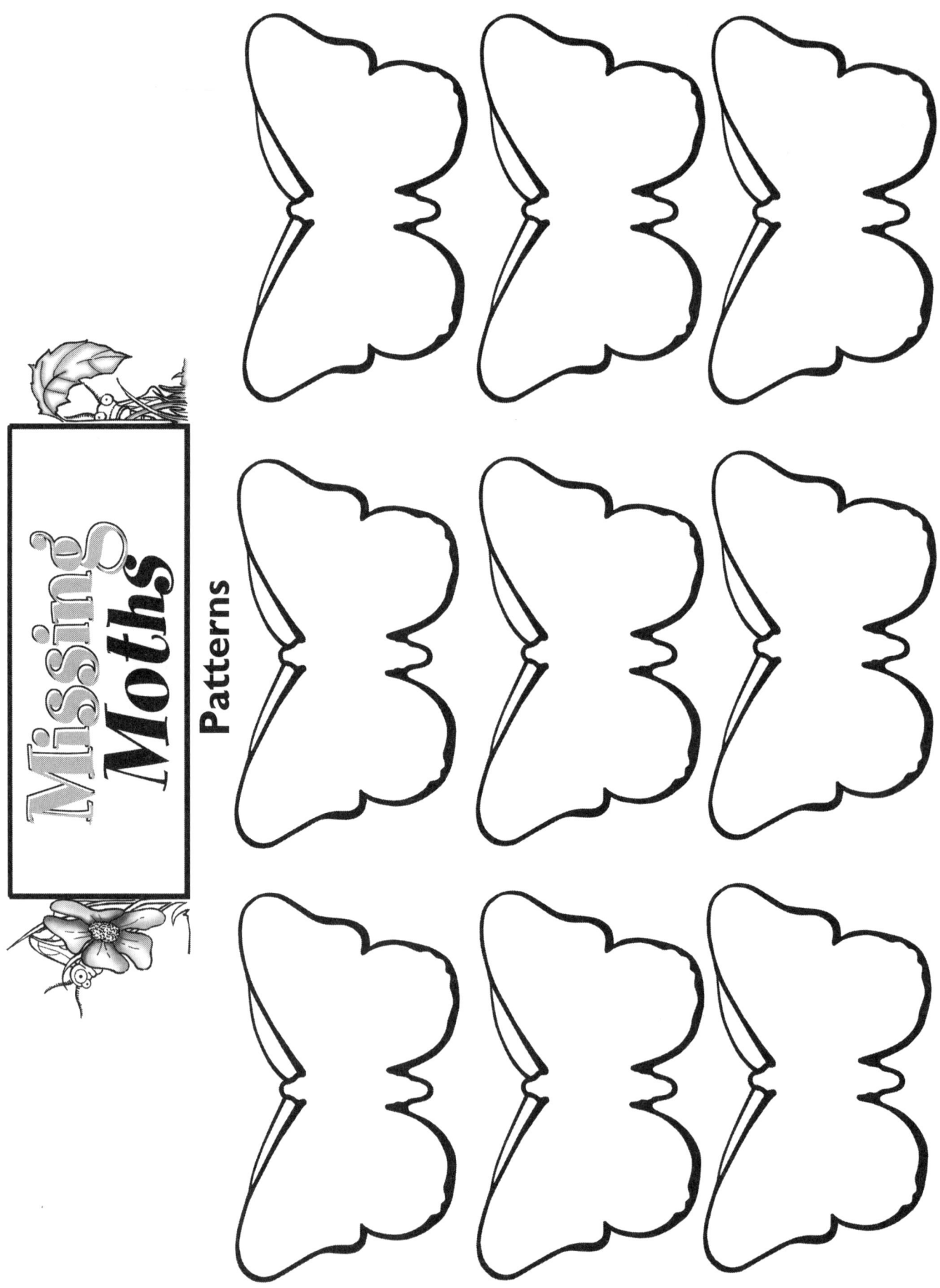

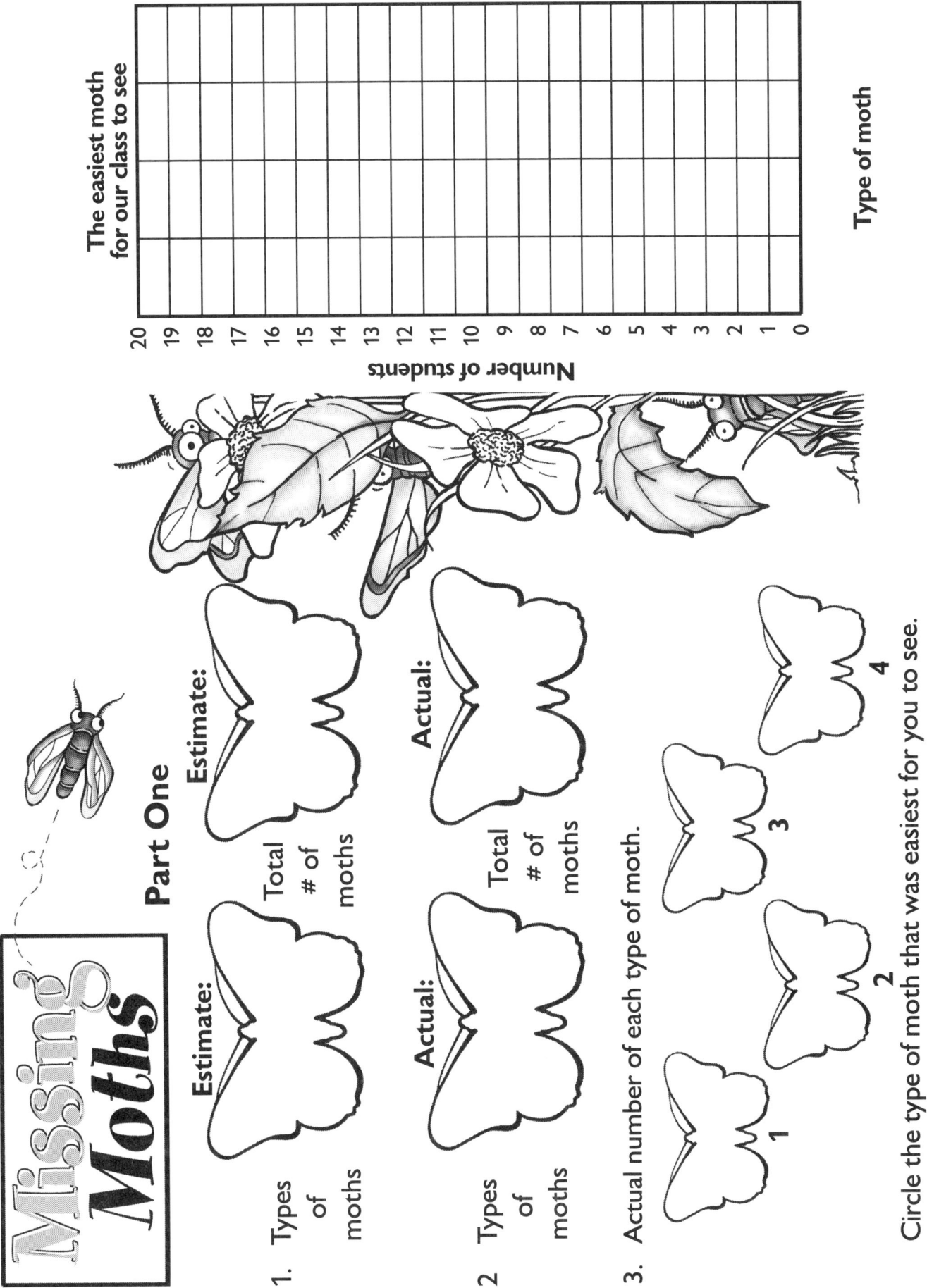

Missing Moths
Part Two

Using the information from page one, complete the table and graph.

Moth type	Estimate	Actual	Estimate/Actual	Decimal equivalent	Percent estimated

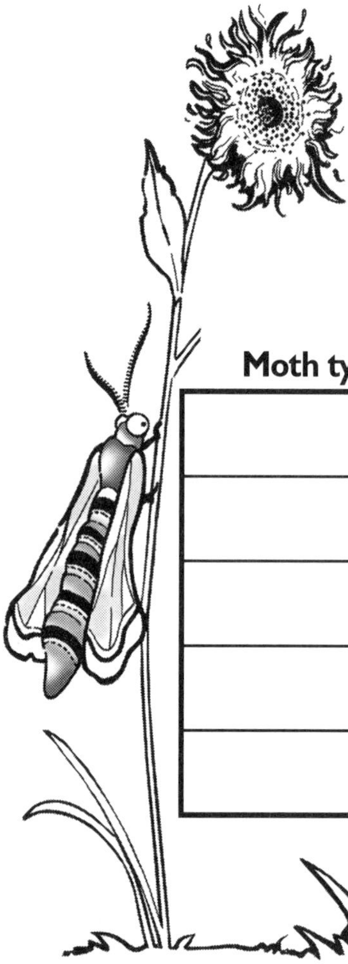

Percent Estimated

Moth type   0   10   20   30   40   50   60   70   80   90   100

Missing Moths

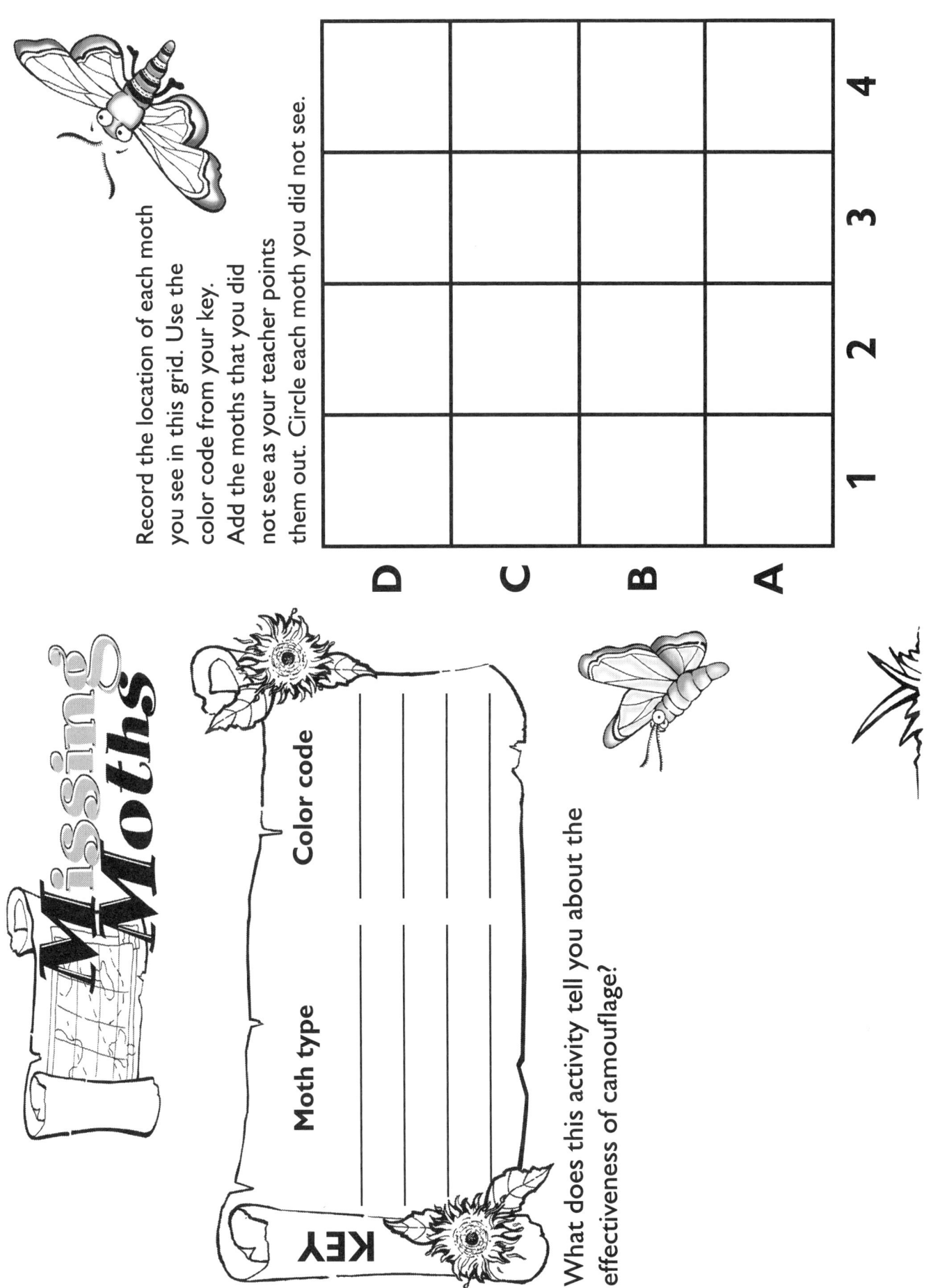

Record the location of each moth you see in this grid. Use the color code from your key. Add the moths that you did not see as your teacher points them out. Circle each moth you did not see.

KEY

Moth type	Color code

What does this activity tell you about the effectiveness of camouflage?

Table Manners

Topic
Insect mouthparts

Key Question
How are insects adapted to eating certain types of food?

Learning Goals
Students will:
1. simulate food gathering with four different types of insect mouths, and
2. determine which mouth is most effective for different food sources.

Guiding Documents
Project 2061 Benchmarks
- *Different plants and animals have external features that help them thrive in different kinds of places.*
- *A model of something is different from the real thing but can be used to learn something about the real thing.*

NRC Standard
- *Each plant or animal has different structures that serve different functions in growth, survival, and reproduction. For example, humans have distinct body structures for walking, holding, seeing, and talking.*

NCTM Standards 2000*
- *Select and apply appropriate standard units and tools to measure length, area, volume, weight, time, temperature, and the size of angles*
- *Collect data using observations, surveys, and experiments*
- *Represent data using tables and graphs such as line plots, bar graphs, and line graphs*

Math
Measurement
   volume
Graphing

Science
Life science
   insect mouthparts

Integrated Processes
Observing
Predicting
Collecting data
Classifying

Materials
Mouthparts (per group of four):
   1 flexible drinking straw
   1 regular straw with one end cut diagonally to form a point
   1 clothespin
   1 clothespin attached to a small piece of sponge about 1" x 1"

Food sources (per group of four):
   paper torn and crumpled into small pieces
   1 narrow-necked bottle containing water
   1 cup containing water and covered with a paper towel or plastic wrap taped securely over its opening
   1 tuna can or small bowl containing water

Collection apparatus (per group of four):
   4 plastic or paper cups

Measuring devices (per group of four):
   graduated cylinders or measuring cups, optional

Background Information
   An adaptation is any characteristic that helps an organism survive. The adaptation may be in the organism's physical appearance, the way its body functions, or the way it behaves. These changes occur through interaction with living and non-living things in the environment.
   An insect's mouthparts are a set of structures used for eating. They surround the insect's mouth. There are two basic types of insect mouths and mouthparts—those adapted for chewing and those adapted for sucking. Many insects have variations or combinations of the two basic types of mouths. For example, the mosquito has a piercing-sucking mouth, a housefly has a sponging mouth, and wasps and bees have chewing-lapping mouths.
   Chewing insects have two grinding jaws called mandibles. They move sideways and are lined with teeth in most species. The jaws are also used for cutting or tearing off food. They have a second pair of less powerful jaws, called maxillae, behind the

mandibles that are used to push the food down the throat. They also have two lips or flaps that hang down over the mouthparts and cover the front of the mouth. The upper lip is called the labrum and the lower lip, the labium. Some examples of chewing insects are grasshoppers, crickets, beetles, cockroaches, and termites.

Sucking insects have mouthparts adapted from the basic chewing mouth structure to suit their feeding habits. The labium in some insects has become a long, grooved beak with four slender, sharp needles called stylets. Stylets are used for piercing and then sucking up juices or blood. In butterflies and moths, the mandibles have lengthened forming a long drinking tube called a proboscis. This tube coils up when the insect is not using it to gather liquids. The mandibles in horseflies have become curved swords that can slash an animal's skin and its maxillae have developed into sharp-pointed rods that can penetrate and extract fluids from the skin.

Management
1. This activity can be done in one of two ways. The first activity sheet has students simulating the different types of insect mouthparts and how they work. It is a good activity to do by itself and is appropriate for primary students. It could also be done as an introductory lesson for the second activity sheet. The second activity sheet allows students to do some measuring and graphing. The set up for each lesson is identical except for the inclusion of measuring cups or graduated cylinders in the second activity to quantify the amount of food collected by the mouthparts.
2. This activity is designed to be done in groups of four.
3. Prepare the materials for each group ahead of time. Each group will need the materials listed.
4. To make management easier, each student should pick one food source and collecting cup to keep for the entire activity. Each student will then use each of the four mouthparts, in turn, to try and collect the "food." It will become obvious that not all the mouthparts work well with all the food sources. For example, the chewing mouth (clothespin) will not work well on anything but the bits of paper, while the straws will not be able to collect the bits of paper. It is important for the students to discuss their experiences with each of the four mouthparts and come to a group consensus as to which mouth is adapted best to each food supply.

Procedure
1. Discuss differences between insects. Focus on methods of eating, different types of mouthparts and various types of food sources.
2. Talk about how different insects have adapted or changed over time to meet environmental changes, food sources, etc.
3. Have students share the different types of mouthparts on various insects that they have observed.
4. Hand out the first activity sheet. Show each type of mouth and discuss how it works. Show the students the four kinds of food sources.
5. Using a cup of water, demonstrate how to capture liquid in a straw with your finger. Note: Caution students not to use their mouths to suck up liquid in the straw. To capture liquid in the straw, lower it into the liquid, and place your finger on top of the straw, trapping the liquid inside. To release the liquid, lower the straw into a "collecting cup," and remove your finger from the top of straw, releasing the liquid. Explain that the piercing-sucking mouthpart (the pointed straw) is the only one that should be used to break through the paper towel or plastic wrap on the covered cup.
6. Hand out a small collecting cup to each student to collect the food gathered by each mouthpart. (All students will use all four mouthparts during the lesson, but will keep the same collecting cup and food supply.)
7. Discuss the activity sheet, and have students fill in their predictions as to which mouth is best suited to each food supply. Explain the mechanics of the lesson. Each "feeding period" should be about two minutes.
8. After students have tried each of the four mouthparts to collect their food, have them discuss their observations and come to a group consensus as to which mouth is best adapted to each food source. They should then fill in the rest of the activity sheet from their observations.
9. The activity can be repeated at a later time using the second activity sheet and having the students count and measure the amounts of each food that the mouthparts collect.

Connecting Learning
1. How does an insect's mouth affect its choice of food?
2. What would happen if all insects had the same type of mouthparts?
3. Where would you look for an insect that had a "sucking" type of mouth? ...chewing? ...piercing/sucking? ...lapping?
4. What are you wondering now?

Extensions
1. Make a list of insects that have each type of mouthpart. Which is the most common type?
2. Discuss how mouthparts relate to where an insect lives.

3. Look up insects in the encyclopedia and draw and label the different types of mouthparts.
4. Design insect mouthparts that would be good at collecting a common food item, such as sugar or fruit.

Curriculum Correlation

Literature
Berger, Melvin and Gilda. *How Do Flies Walk Upside Down? Questions and Answers About Insects.* Scholastic, Inc. New York. 1999.

Mound, Laurence. *Eyewitness Books: Insect.* DK Publishing. New York. 2000.

Wangberg, James K. *Do Bees Sneeze? And Other Questions Kids Ask About Insects.* Fulcrum Publishing. Golden, CO. 1997.

Science
What kinds of adaptations for eating do mammals have? What about reptiles, birds, amphibians, and fish?

Math
Test each mouthpart again allowing a feeding period that is twice as long. Did you eat twice as much?

Geography
Identify as many insects in your area as possible. Are there more insects with one particular kind of mouth? Identify the food source for each insect.

* Reprinted with permission from *Principles and Standards for School Mathematics*, 2000 by the National Council of Teachers of Mathematics. All rights reserved.

Table Manners

Which mouthpart is best adapted to each food source?

Before you begin, predict which mouthpart will be best adapted to each food source. Record your predictions in the table below.

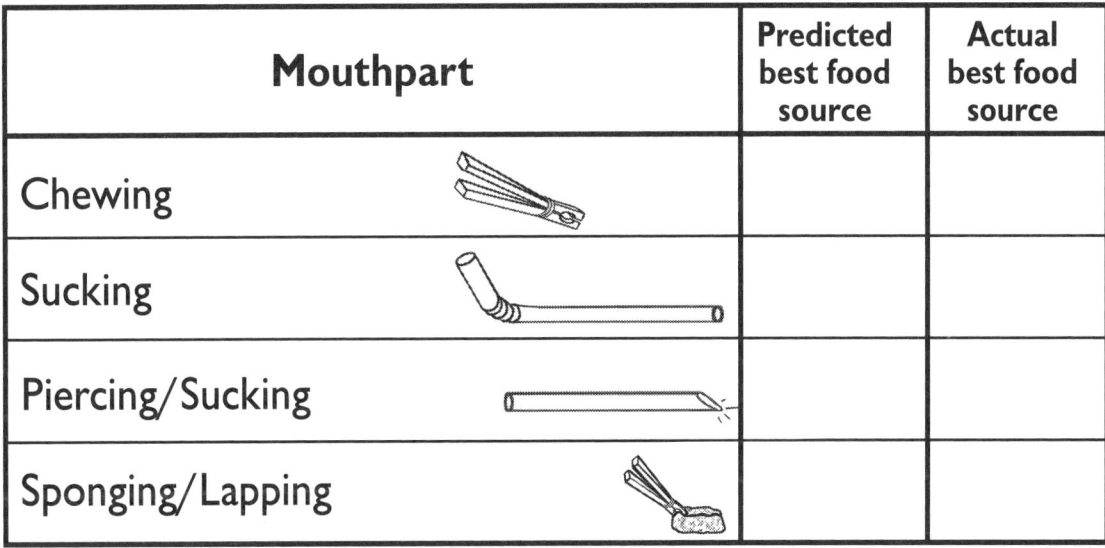

Food Sources

A.

B.

C.

D.

Were your predictions correct? Why or why not?

Match the mouthpart with the critter.

_____   _____   _____   _____

How are these critters adapted to their food sources?

CRITTERS

Table Manners

Which mouthpart will gather the most food?

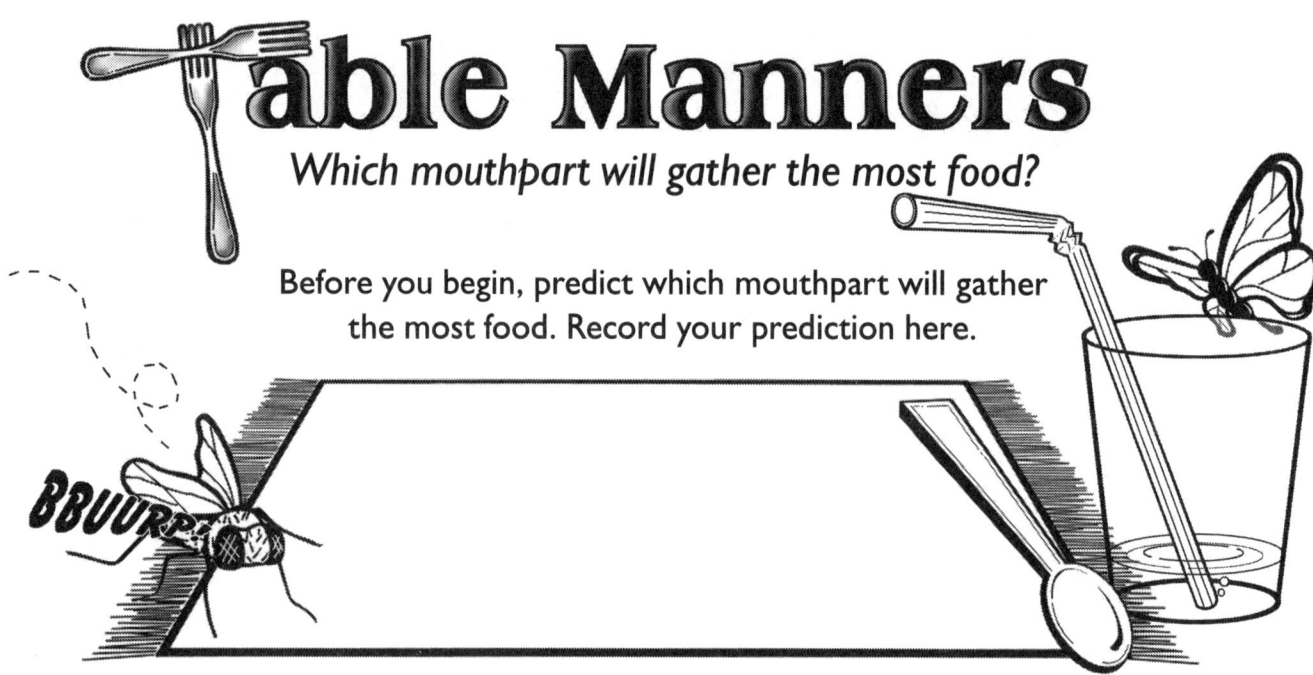

Before you begin, predict which mouthpart will gather the most food. Record your prediction here.

Record the amount of food eaten with each mouthpart at the end of the feeding session.

Mouthpart	Total food eaten
Chewing	pieces
Sucking	mL
Piercing/Sucking	mL
Sponging/Lapping	mL

1. Which mouthpart did you find the easiest to use? Why?

2. Which mouthpart was the most difficult to use? Why?

3. Which mouthpart is best suited to each kind of food source? Justify your response.

CRITTERS

Table Manners

Make a bar graph showing how much food was collected using each of the mouthparts. Be sure to label the blank side of each graph with the appropriate numbers.

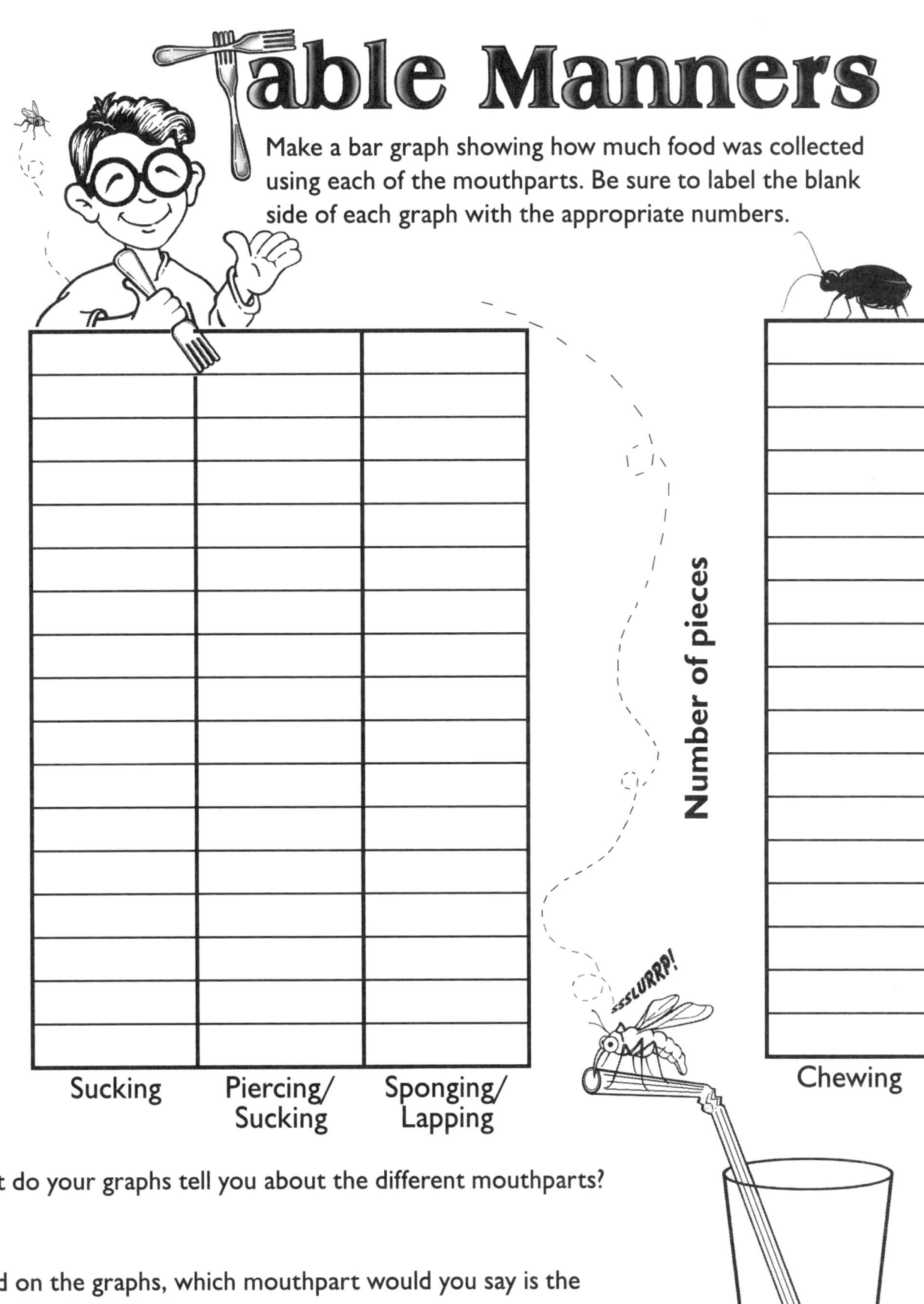

mL of liquid

Sucking | Piercing/Sucking | Sponging/Lapping

Number of pieces

Chewing

What do your graphs tell you about the different mouthparts?

Based on the graphs, which mouthpart would you say is the most effective? Why?

CRITTERS © 2004 AIMS Education Foundation

I'm Stuck on You

Topic
Adaptations

Key Questions
1. How are the tongues of frogs, toads, and some species of lizards adapted to allow them to catch their food?
2. How many times can you successfully catch an insect in 10 trials?
3. Where can you catch the most insects—on the floor, on the wall, or when they are hanging in the air?

Learning Goals
Students will:
1. learn about frog, toad, and chameleon tongues;
2. make a "sticky tongue;" and
3. simulate how some frogs, toads, and chameleons catch their food.

Guiding Documents
Project 2061 Benchmarks
- *In doing science, it is often helpful to work with a team and to share findings with others. All team members should reach their own individual conclusions, however, about what the findings mean.*
- *Plants and animals have features that help them live in different environments.*
- *A model of something is different from the real thing but can be used to learn something about the real thing.*
- *Make something out of paper, cardboard, wood, plastic, metal, or existing objects that can actually be used to perform a task.*

NCTM Standard 2000*
- *Represent data using concrete objects, pictures, and graphs*

Math
Number sense
Graphing

Science
Life science
   adaptations

Integrated Processes
Observing
Predicting
Comparing and contrasting
Generalizing
Applying

Materials
For the class:
   chart paper
   string
   tape
   pictures of insects
   large paper tree, optional (see *Management 5*)
   class graph (see *Management 8*)
   Velcro™ strips or dots
   The Wide Mouth Frog (see *Curriculum Correlation*)

For each student:
   1 party blower (see *Management 1*)
   2 squares of Velcro™ (see *Management 2*)
   1 paper insect

Background Information
The tongues of frogs, toads, and chameleons are wonderful examples of how adaptations make animals better able to survive in their environments. Many frogs, toads, and chameleons have long, sticky tongues that enable them to catch food. The tongues of frogs and toads are covered with slimy mucus to which their prey adheres. Their tongues can extend far out of their mouths to catch insects. Once they have pulled an insect into their mouths, they swallow their meal whole. A frog blinks as it swallows. Its eyeballs help force the food down.

A chameleon's tongue stays bunched up inside its mouth until it is time to catch a grasshopper or other insect. When the tongue shoots out, it is as long as the total length of the chameleon's body and tail. The insect sticks to the club-like padded tip. The tongue and insect are then reeled back into the mouth of the chameleon.

Management
1. Purchase the small-sized party blowers. Do not buy party blowers that make noises or have extra decorations on them which may distract many students. Plain party blowers work best.
2. The Velcro™ can be purchased in squares or in long strips by the yard. If you purchase the strips, cut one-inch squares prior to the activity. Purchase the Velcro™ that has a sticky backing to save time in preparing the "sticky tongues."
3. Designate three different areas of the room that will be used for hunting insects. One area will be on the floor (ground). Another area will be for insects that are hanging in midair (flying), and the third area will be for insects that are on a wall (tree).

CRITTERS

4. Duplicate and cut out enough pictures of insects for each student to have one. Duplicate three extra sets of at least 10 each to use at the stations. You will need to stick the piece of fuzzy-sided Velcro™ onto the center of each insect for the extra sets.
5. To set up the first station, scatter one set of the grasshoppers over a small area on the floor for a group of two or four students to hunt ground insects. For the second station, glue or tape a set of flies to the ends of string and hang them from the ceiling. They should hang no higher than the eye level of students.

For the third station, tape a set of beetles to the wall to simulate insects that are on the sides of trees, bushes, buildings, etc. You can make a large paper tree to mount on the wall to lend an air of realism to this station.

6. Assign partners or groups of students to each area so that everyone is not at the same area at one time.
7. This activity is divided into three parts. The first part involves students in the exploration of party blowers and the construction of the "sticky tongues." The second part has paired students seeing how many insects they can catch in 10 trials. This part is only to provide students with practice in catching insects, and recording and tabulating data. The third part has students rotating through stations to determine which is the easiest location for catching insects: on the ground (floor), in midair, or on a tree (wall).

8. Make a class graph on chart paper or butcher paper. Label the three columns: *On the Ground*, *In the Air*, and *On a Tree*.
9. Before making the sticky tongues, hand out the party blowers and allow children a few minutes to practice blowing. This free exploration is a must! Be sure that students are not too close to other students to avoid hitting each other in the eyes with the party blowers.

Procedure
Part One
1. Read *The Wide Mouth Frog* to the students and then lead them in a discussion about what frogs eat and how they might catch their food. Record their thoughts on large chart paper.
2. Take the discussion further with reference to reptiles such as lizards and other amphibians such as toads by asking *Key Question 1*: How are the tongues of frogs, toads, and some species of lizards adapted to allow them to catch their food?
3. Read some of your own resources to the students or share from the literature connections included with this activity. Discuss why frogs, toads, and some lizards need long, sticky tongues.
4. Tell the students that they are going to pretend to be one of these animals and will be making a sticky tongue that will allow them to catch an insect in a way similar to the animals they have been learning about.
5. Distribute the party blowers. Allow ample time for exploration. This may be a bit chaotic at first but well worth the time spent so you can get their attention for the real purpose of the activity!
6. Caution the students to keep a distance from their classmates to avoid hitting each other in the eyes.
7. Once the students have finished exploring their party blowers, lead them through a discussion about how these blowers are very much like the tongues of frogs, toads, and some lizards.
8. Distribute a square of the looped portion of the Velcro™ to each student.
9. Direct them to blow their party blowers so they are extended out as far as possible. They will need to keep them in this position.
10. Have a partner peel the back off the rough piece of Velcro™ to reveal its sticky surface and stick it on the **underside** of the end of the party blower.

CRITTERS 159 © 2004 AIMS Education Foundation

11. Instruct the students to let the party blowers roll back up and put them aside. Have them help their partners affix the Velcro™ in the same way.
12. Allow each student to choose an insect and color it.
13. Distribute the companion pieces of Velcro™ (smooth/fuzzy), and show them how to affix the Velcro™ to the center of their insects.

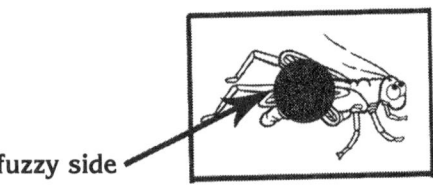
fuzzy side

14. Direct students to place their insects on the floor. Have them blow out their "sticky tongues" and catch their insects.
15. After catching an insect, have students peel them off their "tongues" and continue practicing.

Part Two
1. Ask *Key Question 2:* How many times can you successfully catch an insect in 10 trials?
2. Have students work in pairs. Hand out one copy of the first student sheet to each pair of students. Tell them that one student (*the animal*) will try to catch the insect and the other student (*the scientist*) will keep a record of the trials on the student sheet.
3. Instruct the students who are the *animals* to guess how many times they can catch an insect in 10 trials and record that number on the first student sheet.
4. Ask the *animals* to place their insects on the table and make 10 attempts to catch the insects with their tongues.
5. Each time the *animals* attempt to catch an insect, have the *scientists* make a check mark in the appropriate column on the student page.
6. After all 10 trials, have the *scientist* count up and record the total number of successful catches.
7. Instruct the *animals* and *scientists* to switch roles and repeat this procedure.

Part Three
1. Ask *Key Question 3:* What is the most successful place to catch an insect—on the floor, on a wall, or hanging in the air?
2. Identify the three insect-collection areas already set up in the room.
3. Have students work with partners or small groups.
4. Give instructions for how you would like students to rotate through the insect-collection areas and distribute the second student sheet to each student.
5. Tell students that in each area they will take turns trying to catch the insects 10 different times. Each person must record his or her own data on the recording page.
6. After all students have been through the three stations, distribute the *Pest Strips* bar graphs. On each half of the page, have each student color the insects on the graph to represent his or her successful catches in *Part Three*.
7. Once students have recorded their results, have them cut out the strips on the right half of the page to put on a large class graph.
8. Combine the individual results into a class graph to see in which area the students were most successful.
9. Close with a final time of class discussion and sharing.

Connecting Learning
1. What techniques did you use to catch the insects? [Answers will vary but may include: a fast flick of the tongue, a slow flick of the tongue; being close to the insect, standing as far from the insect as possible]
2. Which technique worked the best for you? Why do you say that?
3. Was anyone successful in catching an insect 10 times in a row?
4. Do you think frogs, toads, and chameleons catch their meals every time they try?
5. In which hunting place was it easiest for you to catch your insects? Why?
6. Where do you think it is easiest for frogs, toads, and chameleons to catch their food? Why?
7. Was your actual catch in *Part Two* close to what you had guessed? Why or why not?
8. Do you think your easiest catch was the easiest for the rest of the class? Explain.
9. What are you wondering now?

Extensions
1. Let students select an animal (frog, toad, or chameleon) and design a mask with a hole where the tongue would be. They will need to make the hole big enough for the party blower to fit through. The students will also need to cut two holes for eyes so they can find their insects!
2. Discuss the roles of predator and prey in this activity. [Frog is the predator; insect is the prey.] Then ask the students when an insect can be a predator and when a frog can be the prey. Have them name animals that would be predators of frogs. [birds, snakes, etc.]
3. Have students design a possible food chain that would include the insect and either a frog, toad, or chameleon. For example: Bear—> raccoon—> fish—> frog—> insect—> leaves.

Internet Connections
Frog adaptations:
http://www.exploratorium.edu/frogs/mainstory/frogstory2.html

Chameleon adaptations:
http://www.pbs.org/edens/madagascar/creature3.htm

Curriculum Correlation
Bennett, Paul. *Catching a Meal*. Thomson Learning. New York. 1994.

Carle, Eric. *The Mixed-Up Chameleon*. Harper Trophy. New York. 1988.

Clarke, Barry. *Eyewitness Books: Amphibian*. DK Publishing. New York. 2000.

Downer, John. *Weird Nature: An Astonishing Exploration of Nature's Strangest Behavior*. Firefly Books. Buffalo, NY. 2002.

Frogs (Face-to-Face). Scholastic, Inc. New York. 2001.

Gibbons, Gail. *Frogs*. Holiday House. New York. 1994.

Lovett, Sarah. *Extremely Weird Frogs*. Avalon Travel Publishing, John Muir. New York. 1996.

Pallotta, Jerry. *The Yucky Reptile Alphabet Book*. Charlesbridge Publishing. Boston. 1990.

Parsons, Harry. *The Nature of Frogs: Amphibians with Attitude*. GreyStone Books. Berkley, CA. 2000.

Schneider, Rex. *The Wide-Mouthed Frog*. Stemmer House. Baltimore, MD. 1991.

* Reprinted with permission from *Principles and Standards for School Mathematics*, 2000 by the National Council of Teachers of Mathematics. All rights reserved.

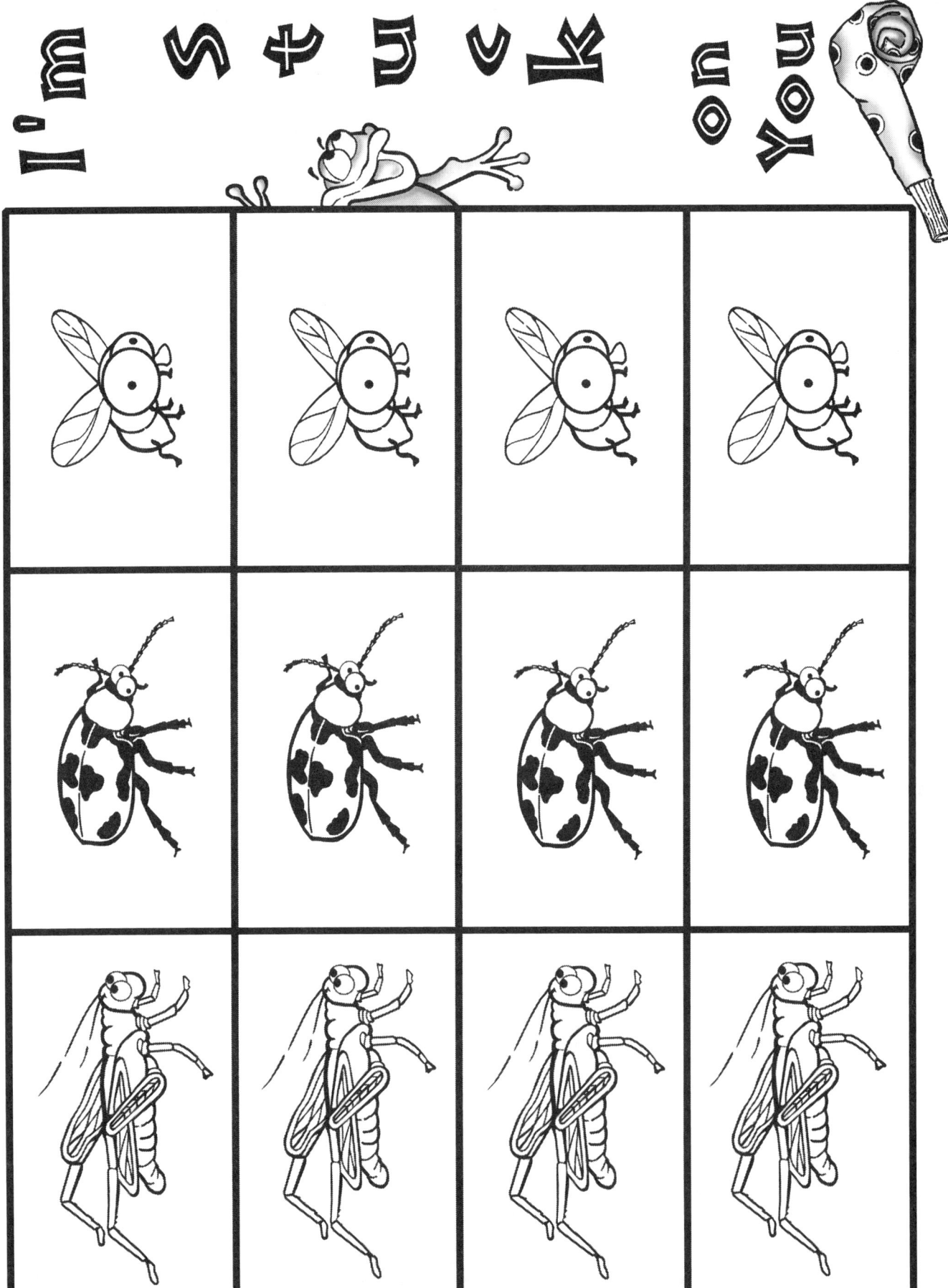

I'm Stuck on You

Animer _____ Scientist _____
I think I will catch ____ insects.

Trial	Caught	Missed
1		
2		
3		
4		
5		
6		
7		
8		
9		
10		

I actually caught ____ insects.

Animal _____ Scientist _____
I think I will catch ____ insects.

Trial	Caught	Missed
1		
2		
3		
4		
5		
6		
7		
8		
9		
10		

I actually caught ____ insects.

CRITTERS © 2004 AIMS Education Foundation

Pest Strips

 Record the number of times you caught each bug by coloring in that many squares. Color both sets of strips.

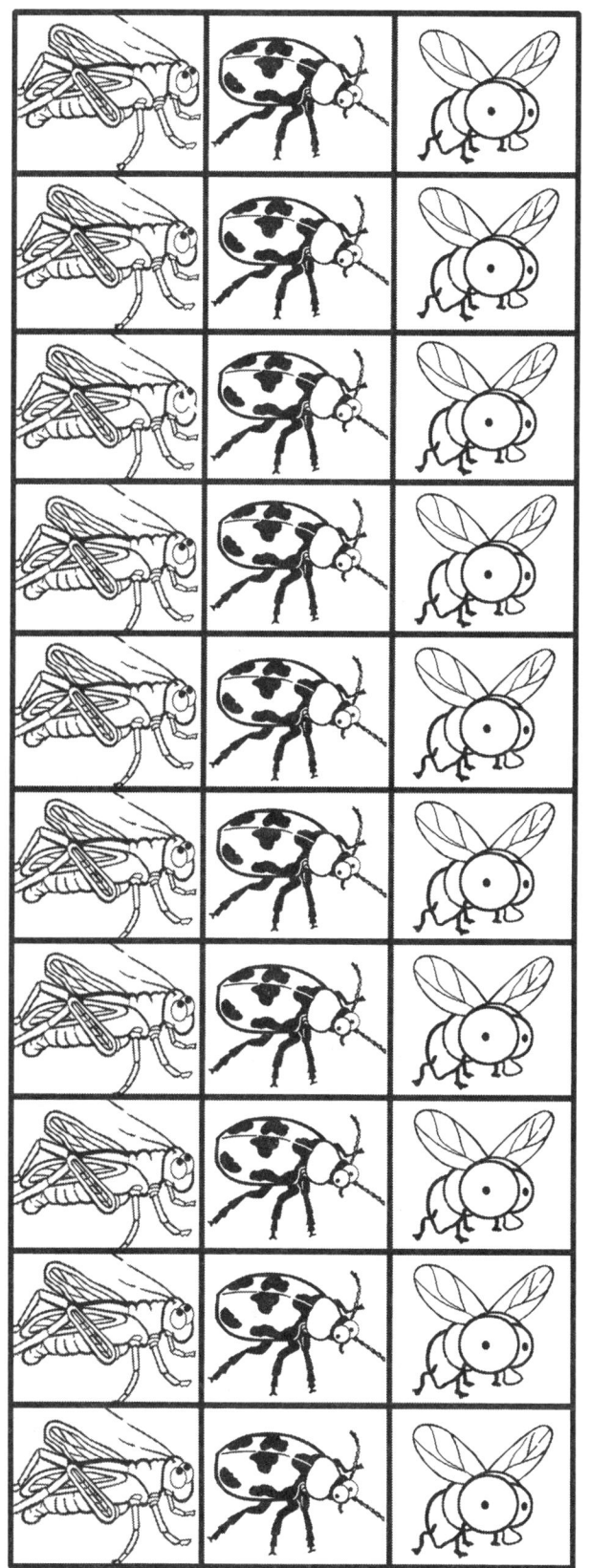
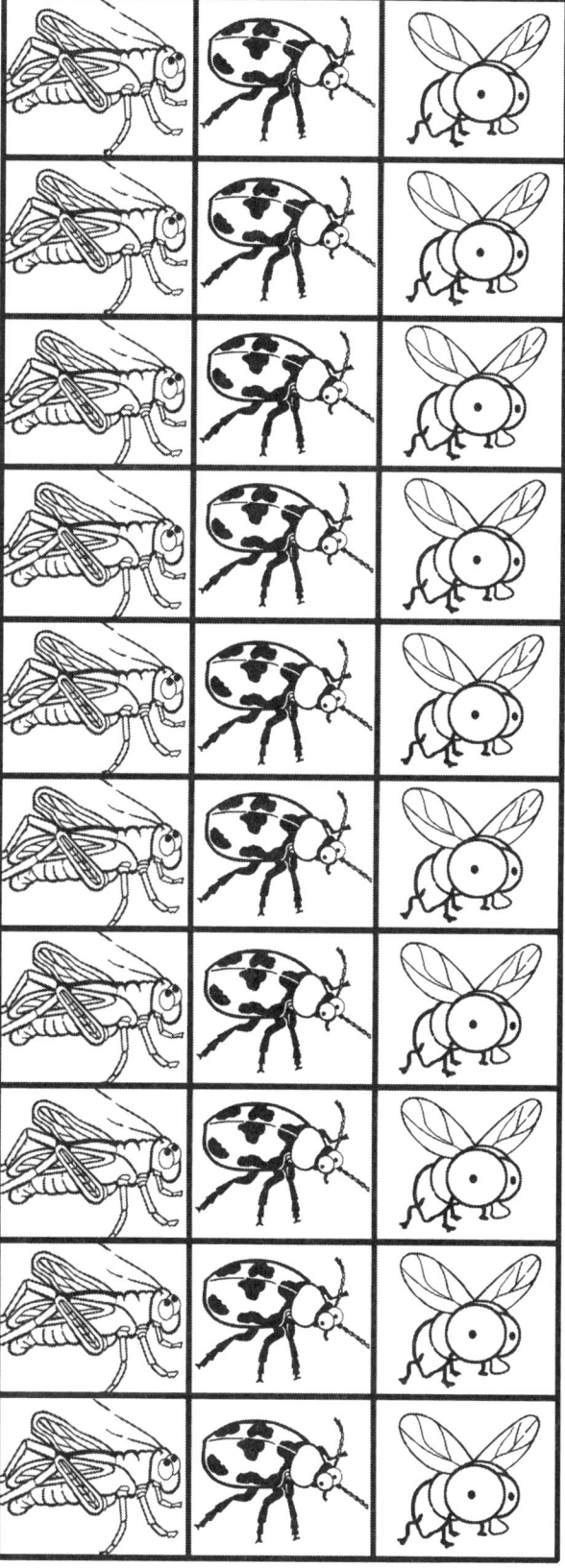

CRITTERS © 2004 AIMS Education Foundation

Mealworms on Stage

Topic
Mealworm populations

Key Question
How will a mealworm population change over a period of four weeks?

Learning Goal
Students will observe and record data for a changing population of mealworms.

Guiding Documents
NRC Standards
- Plants and animals have life cycles that include being born, developing into adults, reproducing, and eventually dying. The details of this life cycle are different for different organisms.
- Ask a question about objects, organisms, and events in the environment.
- Plan and conduct a simple investigation.

*NCTM Standard 2000**
- Represent data using tables and graphs such as line plots, bar graphs, and line graphs

Math
Graphing
Counting

Science
Life science
   animals
      mealworms

Integrated Processes
Observing
Classifying
Manipulating variables
Inferring

Materials
Class culture of mealworms
Both *Mealworms on Stage* activity sheets for each student

Background Information
See the mealworm rubber band book.

Management
1. Buy some mealworms from a bait shop, pet store, or one of the sources listed in *Resources*. Begin a class culture about two weeks before you collect the first data. To make a classroom culture, you will need mealworms, oatmeal or bran, a plastic container, and a potato. Add the bran or oatmeal to the container to form a layer two to three inches deep. Put in a supply of mealworms and both halves of a potato. A cover is not necessary.
2. Make a large copy of the data table and post it on a wall above the mealworm culture.
3. Decide on what day of the week you will do your population counts.
4. Make sure that students know how to identify each stage of the mealworm's life cycle.
5. Allow time for the students to observe the mealworms before beginning the population count.
6. Students will need four copies of the first activity sheet (for group results), one for each week.

Procedure
1. Have each group scoop up a portion of the mealworm population to count. Make sure that all the mealworms are taken from the class culture.
2. Ask groups to classify their samples according to life stages, and count the number of mealworms at each stage. Have them record this information on the first page (group results page).
3. Come together as a class and have each group report their data. Instruct all students to fill in the data in the appropriate spaces on the first page.
4. After each group has reported the number of mealworms at each stage, have students add the columns to find the total number of mealworms in the class culture at each stage.
5. Have students record class totals on the second activity sheet and begin the line graph.
6. Repeat these procedures three more times at one-week intervals. Each time students will need a new copy of the first activity sheet to find the group and class totals. The second activity page should be continued each week. At the end of four weeks, the chart and graph will be complete.

Connecting Learning
1. How many mealworms were at each stage in the life cycle?
2. How did these numbers change from the previous observation?

CRITTERS   © 2004 AIMS Education Foundation

3. Are there any patterns in the mealworm's life cycle that can be determined from this activity?
4. Can you determine the length of each stage in the mealworm's life cycle?
5. What are you wondering now?

Extensions
1. Continue the activity for longer than four weeks and see if any long-term patterns in the mealworm's life cycle emerge.
2. Find the mass of the potato each week and explain the changes.
3. Devise a way to accurately determine the length of each stage in the mealworm's life cycle.
4. Make the "Metamorphosis Wheel" depicting the mealworm's life cycle.

Curriculum Correlation
Literature
Mason, Adrienne. *Mealworms: Raise Them, Watch Them, See Them Change*. Kids Can Press. Buffalo, NY. 2001.

Schaffer, Donna. *Mealworms*. Bridgestone Books. Mankato, MN. 1999.

Language Arts
Pretend you are in the pupae stage and tell what it is like to go through metamorphosis.

Math
Convert the numbers on the chart to percentages.

Science
Make a list of other insects that go through a complete metamorphosis.

Resources
If you are not able to find mealworms in a local pet store or bait shop, King mealworms are available from the following sources:

Bassett's Cricket Ranch
1-800-634-2445
http://www.bcrcricket.com

Carolina Biological Supply Company
1-800-334-5551
http://www.carolina.com

Insect Lore
1-800-LIVE BUG
http://www.insectlore.com

Timberline
1-800-423-2248
http://www.timberlinefisheries.com

* Reprinted with permission from *Principles and Standards for School Mathematics*, 2000 by the National Council of Teachers of Mathematics. All rights reserved.

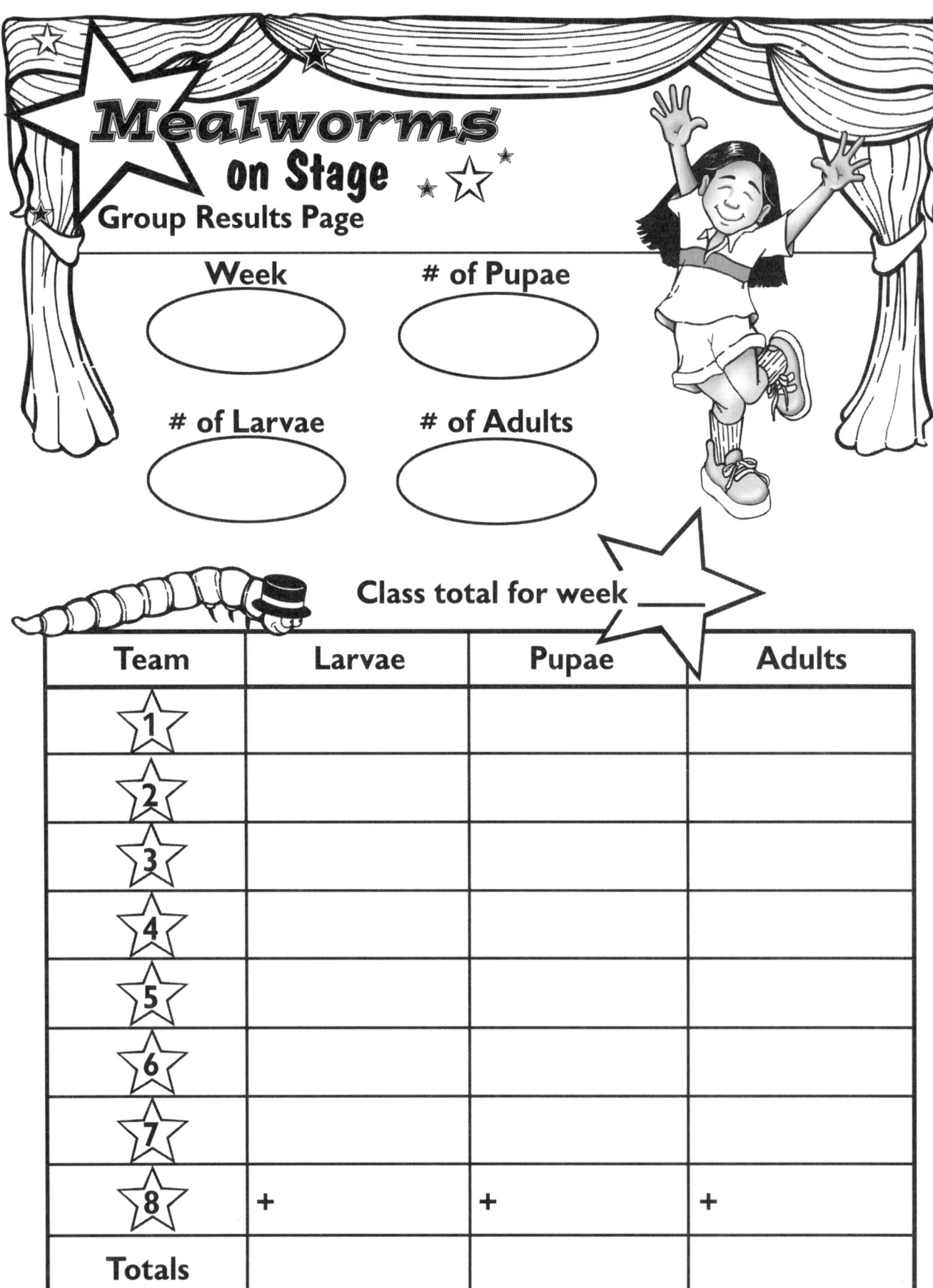

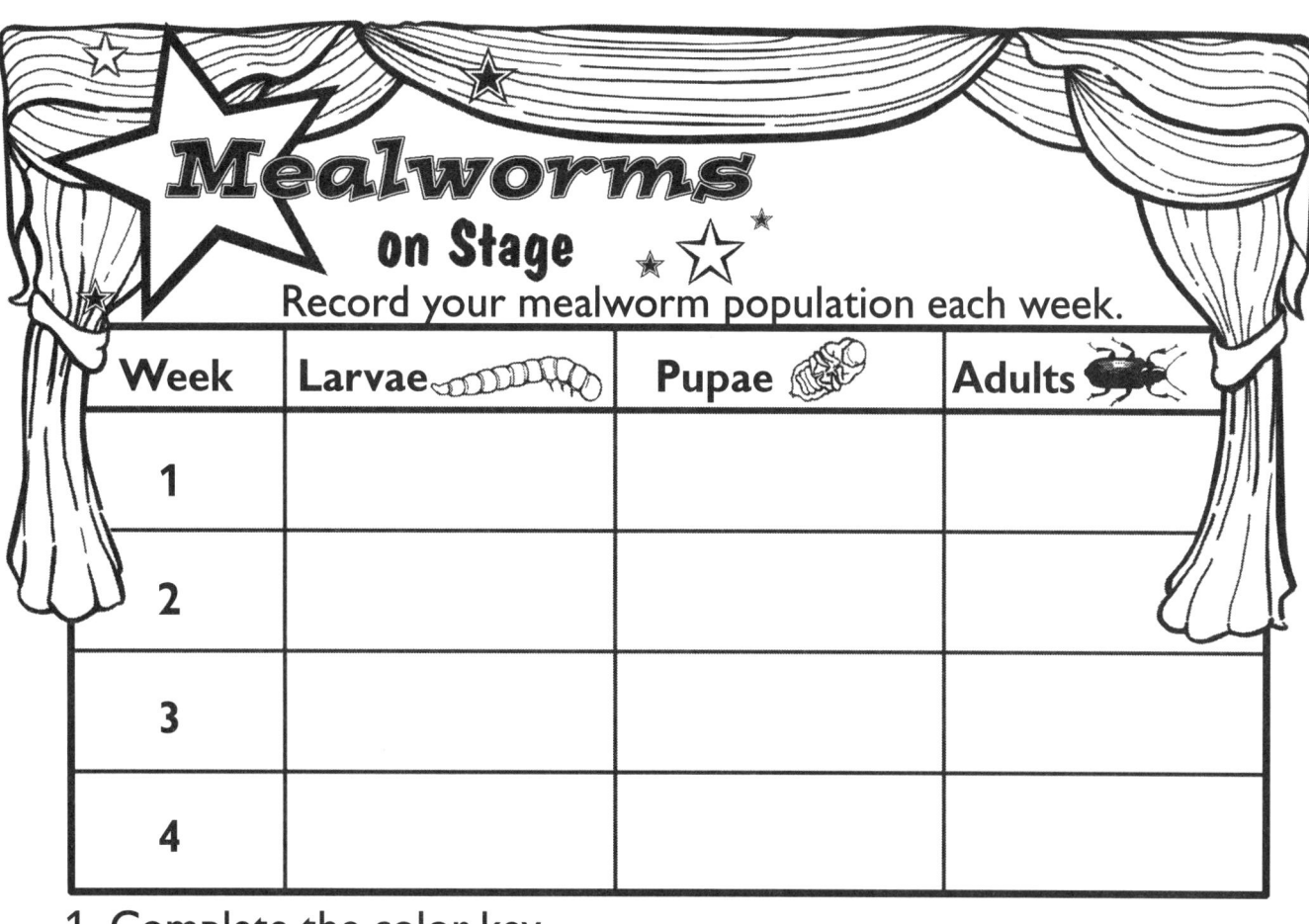

Mealworms on Stage

Record your mealworm population each week.

Week	Larvae	Pupae	Adults
1			
2			
3			
4			

1. Complete the color key.
2. Make a line graph for each stage.

Week 1   Week 2   Week 3   Week 4

Key

CRITTERS

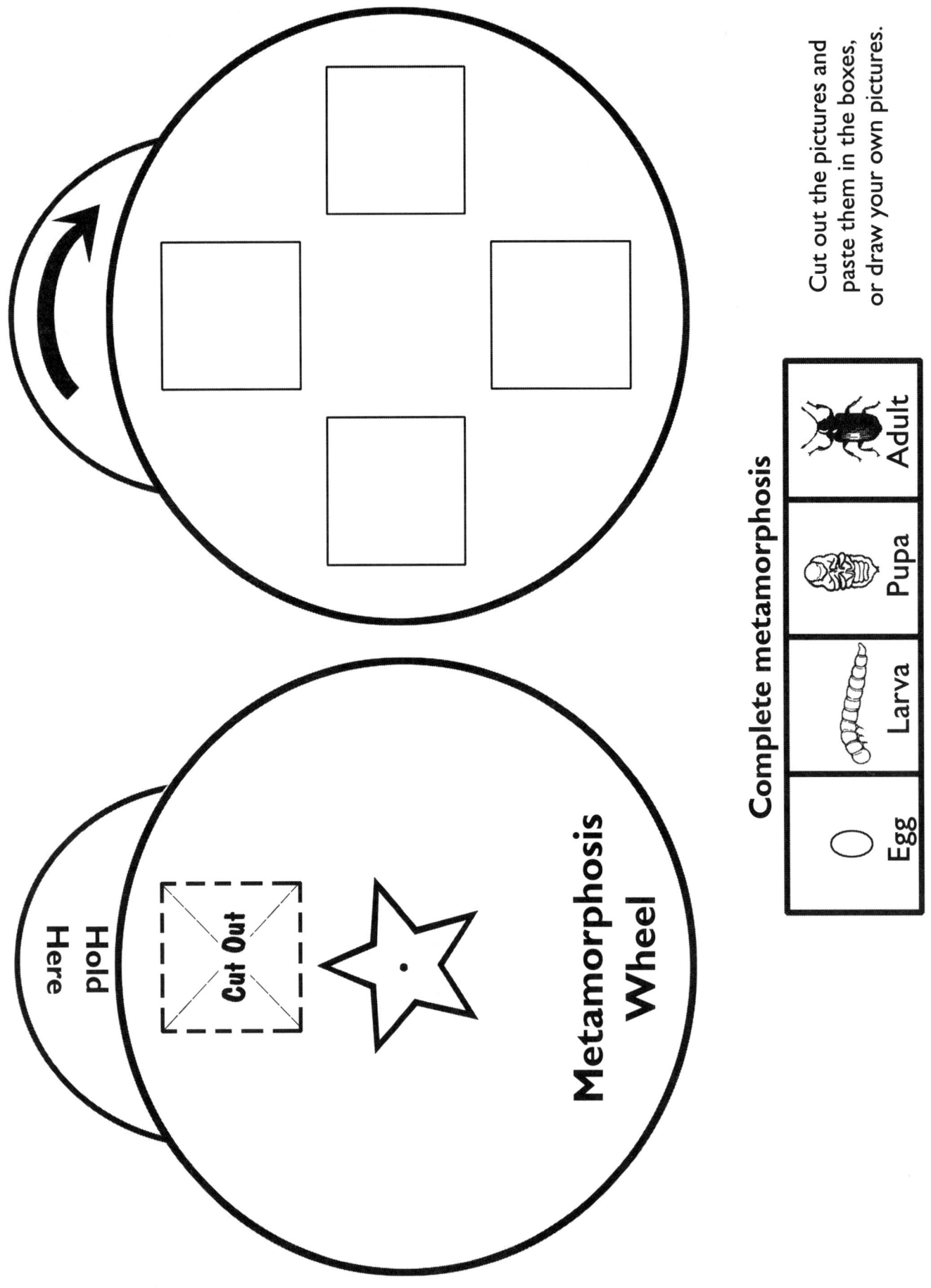

This is Your Life, Tadpole

Topic
Life cycle of frogs and toads

Key Question
How do frogs and toads change as they go through their life cycles?

Learning Goals
Students will:
1. learn how frogs and toads grow from egg to adult;
2. observe and/or research the development of a frog or toad from an egg, to a tadpole, to an adult; and
3. make a model showing the life cycle of a frog or toad.

Guiding Documents
Project 2061 Benchmarks
- Plants and animals have features that help them live in different environments.
- Animals eat plants or other animals for food and many also use plants (or even other animals) for shelter and nesting.
- Buttress their statements with facts found in books, articles, and databases, and identify the sources used and expect others to do the same.

NRC Standards
- Plants and animals have life cycles that include being born, developing into adults, reproducing, and eventually dying. The details of this life cycle are different for different organisms.
- Each plant and animal has different structures that serve different functions in growth, survival, and reproduction. For example, humans have distinct body structures for walking, holding, seeing, and talking.
- Organisms have basic needs. For example, animals need air, water, and food; plants require air, water, nutrients, and light. Organisms can survive only in environments, and distinct environments support the life of different types of organisms.

Science
Life science
  animals
    life cycles
      frogs and toads

Integrated Processes
Observing
Researching
Communicating
Comparing and contrasting
Collecting and recording data

Materials
Part One
  frog or toad eggs and/or tadpoles
    (see *Management 1 & 2*)
  tadpole habitat (see *Collecting and Caring for Tadpoles*)
  Tadpole Observation Journal
  chart paper
  frog and toad research materials
  stapler

Part Two
  plastic drink bottles (20 oz), one per student
    (see *Management 5*)
  transparent tape
  scissors
  green copy paper
  Life Cycle Strips
  Stage Viewer Strips
  colored pencils, optional

Background Information
Most frogs and toads begin their lives as small, soft eggs laid in water. Frogs tend to lay their eggs in large clusters, while toad eggs are usually laid in long strings attached to vegetation. The eggs are round and surrounded by a clear, jelly-like substance that keeps the eggs moist and protects them from disease. Eggs usually hatch anywhere from three to 25 days after being laid. The number of eggs laid can be as few as six to as many as 20,000.

When the eggs hatch, they are called tadpoles—the larval stage of frogs and toads. Tadpoles have large heads and bodies with gills and small tails. They feed mostly on algae and other vegetable matter in the water. As the tadpoles grow larger, their bodies begin to change. First, they grow hind legs, then front legs appear. As their legs grow longer, their tails grow shorter. Their insides are changing too. Their gills are disappearing to be replaced by lungs. Their digestive systems change from being suited to eating plants to being suited for protein. Soon the froglets or baby toads leave the water on all four legs. The last stub of their tails shrinks away, and they are adults.

Management
1. It is ideal if students can observe live tadpoles for this activity. If you live in an area where frogs or toads can be found, you can collect eggs and/or tadpoles from local water sources such as ponds, marshes, or garden pools (see *Collecting and Caring for Tadpoles* for more information).
2. If you do not have access to frog or toad eggs/tadpoles locally, they can be ordered from a variety of biological supply companies. The *Resources* section lists several options. **Caution:** Do research before ordering to determine what frog or toad species are native to your area. It is very difficult to care for adult frogs and toads, so they need to be released to the wild once they reach adulthood. **DO NOT** release non-native frog or toad species into the wild. This can have severe repercussions on the local ecosystem.
3. Collect a variety of books and other sources where students can find pictures and information about the life cycle of frogs and toads. (See *Resources* for suggestions.)
4. Each student will need his or her own *Tadpole Observation Journal*. To assemble the journals, copy the pages front to back as they are in the book. Fold along the center line and nest the pages together so that the page numbers are in the correct order. Staple along the spine of the journal so that none of the pages can come apart.
5. For *Part Two*, each student will need his or her own 20-oz plastic drink bottle. The bottles that have an indentation in the middle for the label work the best.
6. Each student will need one *Life Cycle Strip* and one *Stage Viewer Strip*. The *Stage Viewer Strips* should be copied onto green paper so that they contrast with the white of the *Life Cycle Strips*. You may wish to cut out the rectangular viewing windows for students ahead of time.

Procedure
Part One—With live tadpoles
1. Ask the *Key Question* and discuss what students already know about the life cycle of frogs and toads. Record student responses on a piece of chart paper.
2. Distribute the *Tadpole Observation Journal* pages to each student. Assist them in assembling their journals. Have them record the main points of the class discussion on the first page of the journal.
3. Have students begin by doing some research on frog and toad eggs. They should try to discover what they look like, where they are laid, how long they take to hatch, and so on. If you have eggs in the classroom, provide time for students to observe the eggs. Ask students to record their findings and observations.
4. Over the next few weeks, have students observe the tadpoles carefully and record their observations. For the first week, allow time for observation every day. For the next few weeks, have observation time twice a week.
5. In addition to making observations, students will need to conduct research about the stages in a tadpole's life. Have them record their findings in their journals.
6. The observation journals ask students to research and draw three different changes that a tadpole goes through on its journey to becoming a frog. Students should be allowed to select the changes on which they would like to focus; however, the changes should be recorded in chronological order.
7. Have students complete their journals by summarizing what they learned and writing down any questions they still have about the life cycle of frogs and toads.

Part One—Without live tadpoles
1. Ask the *Key Question* and discuss what students already know about the life cycle of frogs and toads. Record student responses on a piece of chart paper.
2. Distribute the *Tadpole Observation Journal* pages to each student. Assist them in assembling their journals. Have them record the main points of the class discussion on the first page of the journal.
3. Have students begin by doing some research on frog and toad eggs. They should try to discover what they look like, where they are laid, how long they take to hatch, and so on. Ask students to record their findings and make a detailed drawing of frog or toad eggs.
4. Continue this research and recording for each stage of the tadpole's life. The observation journals ask students to research and draw three different changes that a tadpole goes through on its journey to becoming a frog. Students should be allowed to select the changes on which they would like to focus; however, the changes should be recorded in chronological order. In each case, emphasize the importance of finding a good picture to record.
5. Have students complete their journals by summarizing what they learned and writing down any questions they still have about the life cycle of frogs and toads.

Part Two
1. Distribute one bottle and a *Life Cycle Strip* to each student. If desired, allow students to color in the stages of the frog's life. You may also choose to have students make their own strips using the pictures from their observations.

2. Demonstrate how to wrap the *Life Cycle Strip* around the middle of the bottle and tape it securely in place. For right-handed students, the strip should be attached so that the adult frog faces the opening of the bottle. For left-handed students, the adult frog should face the bottom of the bottle.

3. Give each student a *Stage Viewer Strip*. If necessary, have them cut out the rectangular viewing window.
4. Assist students in correctly taping the *Stage Viewer Strip* around the *Life Cycle Strip* so that it spins freely and shows only one stage at a time.
5. Have students spin their bottles and describe the different stages of a frog's life to a classmate.

Connecting Learning
1. What do baby frogs and toads look like? [They are tadpoles, like little fish.] Where are they found? [in the water]
2. Are baby frogs and toads born? [No, they hatch from eggs.]
3. What do you know about the life cycle of frogs and toads?
4. What are the different stages of a tadpole's life?
5. How long did it take for our tadpoles to change into frogs/toads?
6. What was the first change that you saw in our tadpoles?
7. What was the most surprising change that you saw in our tadpoles?
8. What other animals or insects go through metamorphosis?
9. What are you wondering now?

Extensions
1. Devise a series of experiments to answer questions about your tadpoles. Do they like light or dark better? Do they prefer warm water or cool water? Do they like to eat at the top of the water, in the middle, or near the bottom? How fast do they grow?
2. Make and label anatomical drawings of the tadpole at several stages in its life cycle.

Curriculum Correlation
Berger, Melvin and Gilda. *How Do Frogs Swallow With Their Eyes? Questions and Answers About Amphibians.* Scholastic, Inc. New York. 2002.

Gibbons, Gail. *Frogs.* Holiday House. New York. 1993.

Pascoe, Elaine. *Tadpoles.* Blackbirch Press, Inc. Woodbridge, CT. 1997.

Internet Connections
Frogland
http://allaboutfrogs.org/weird/general/cycle.html
This well-done site has a simple description of the life cycle of frogs using kid-friendly language. Also has tips and instructions for raising tadpoles.

Resources
Carolina Biological Supply
2700 York Road
Burlington, NC 27215
1-800-334-5551
http://www.carolina.com
Both frog eggs and live tadpoles available.

Connecticut Valley Biological
82 Valley Road
P.O. Box 326
Southampton, MA 01073
1-800-628-7748
http://www.ctvalleybio.com/
Frog eggs, bullfrog (Rana catesbeiana), and woodfrog (Rana sylvatica) tadpoles available.

Insect Lore
P.O. Box 1535
Shafter, CA 93263
1-800-LIVE BUG
http://www.insectlore.com/store.html
Frog Hatchery Kit includes aquarium, magnifying glass, tadpole food, and a coupon for live frog embryos.

NASCO—Fort Atkinson
901 Janesville Avenue
P.O. Box 901
Fort Atkinson, WI 53538-0901
NASCO—Modesto
4825 Stoddard Road
P.O. Box 3837
Modesto, California 95352-3837
1-800-558-9595
http://www.nascofa.com/
Both frog eggs (Rana pipiens) and live tadpoles (Rana pipiens and Xenopus laevis) at various stages are available.

 # Collecting Eggs and/or Tadpoles

When to Collect
The best time to look for eggs and tadpoles is in the spring when the weather starts getting warmer. Eggs can hatch very quickly, so don't wait too long after the weather starts to warm up or you may miss your opportunity.

Where to Collect
The best place to find eggs and tadpoles is in fresh water that is shallow and fairly still. The edges of ponds and marshes, ponding basins, garden pools, and ditches are all places you might look. Tadpoles will often be found where water plants are growing.

Cautions
Be sure to take appropriate safety precautions when collecting eggs or tadpoles. If you are collecting from a public location such as a park, obtain the proper permission to remove the eggs or tadpoles. Collect only as many eggs and/or tadpoles as you can care for properly. You probably won't want more than a few, or your containers will get too crowded.

 # Caring for Tadpoles

Water
Tadpoles need fresh, pure water. Even the smallest bit of chlorine (added to most tap water) can be deadly. Use bottled spring water or allow tap water to sit out for at least 24 hours before placing the tadpoles in it. You can also buy chlorine remover drops at a pet store that will purify the water. You will need to change the water frequently to keep it fresh and clean. You can replace some or all of the water when you do this. If you clean the container, do not use soap or other chemicals, as any residue could harm the tadpoles.

Container
The best containers for observation are clear plastic aquariums that can be purchased at pet stores. Do not use a container that has held any kind of chemicals, cleaning products, etc. If you have a large number of tadpoles, you may need to use a plastic wading pool to accommodate them all. Do not put a lid over the container except for short periods of time to move it. If the supply of oxygen to the water is cut off, the tadpoles will die.

Number of Tadpoles
Suggestions vary about the number of tadpoles that can be housed in a single container. Crowded conditions aren't good for the tadpoles. You should never have more than about five or six tadpoles per gallon of water.

Food
Tadpoles' food in the wild consists mainly of algae and other plant matter. To replicate this, boil lettuce or spinach leaves for about a minute (do not use cabbage). You can also buy dry food designed for tadpoles at some pet stores. As the tadpoles get older, you may want to add hard-boiled egg yolk to their diet for protein. Remove any uneaten food from the container at the end of each day. This will help keep the water clean.

Location
Be sure the container is at least three-fourths shaded at all times. If the water is exposed to sunlight for too long, it will be heated above room temperature, which is not good for the tadpoles.

Other
Once the tadpoles begin to grow legs, place an object such as a rock or piece of wood in the water. Be sure that it sticks out enough to provide the froglets a place to rest out of the water.

This is Your Life, Tadpole

Life Cycle Strips

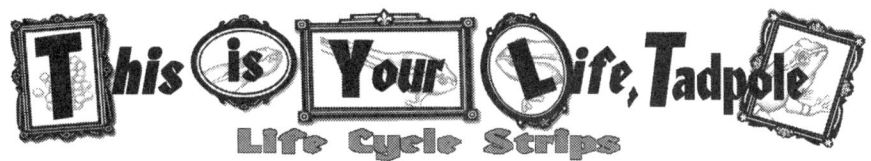

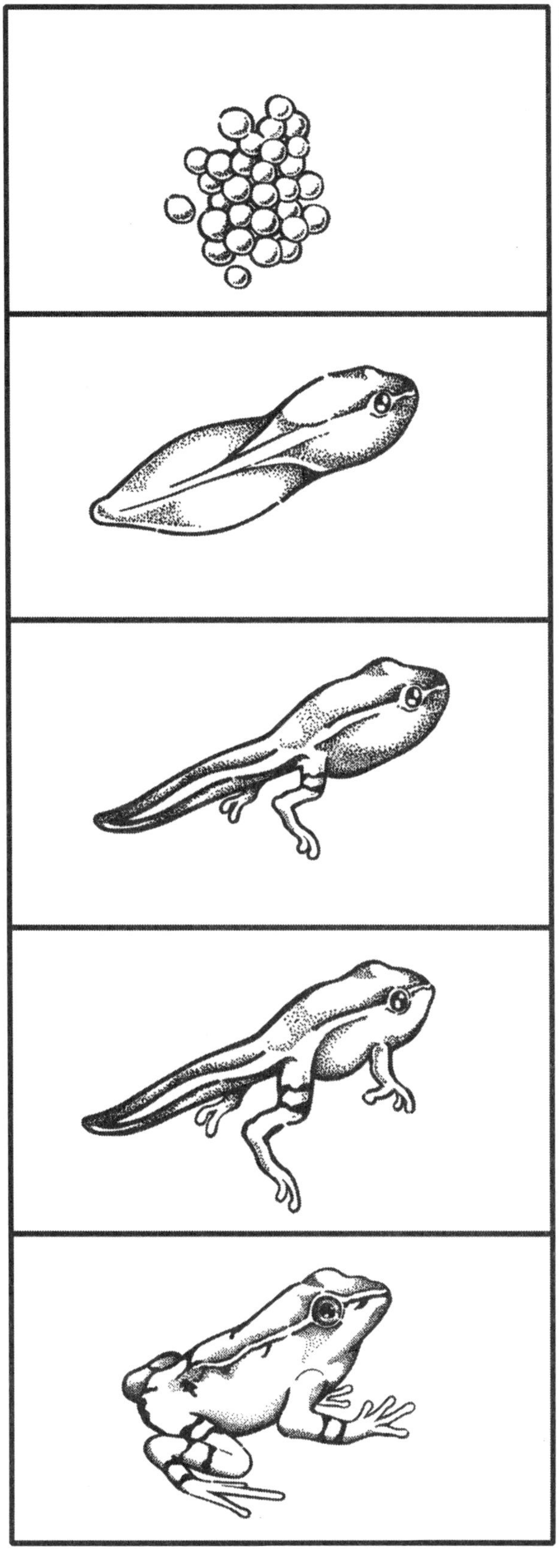

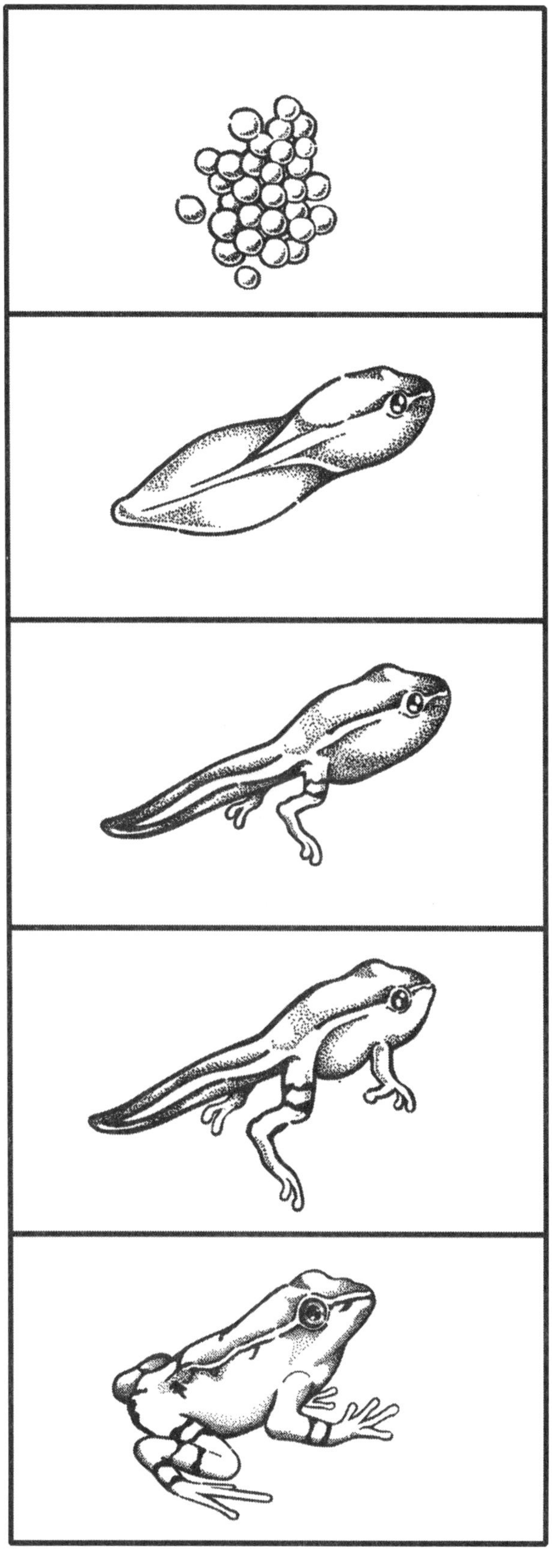

This is Your Life, Tadpole

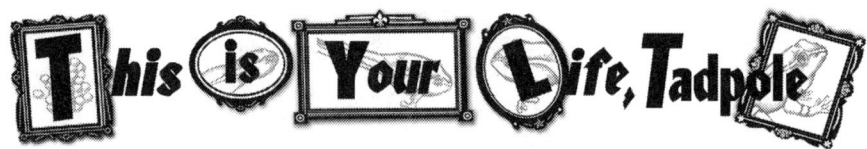

Life Cycle of a Frog

Life Cycle of a Frog

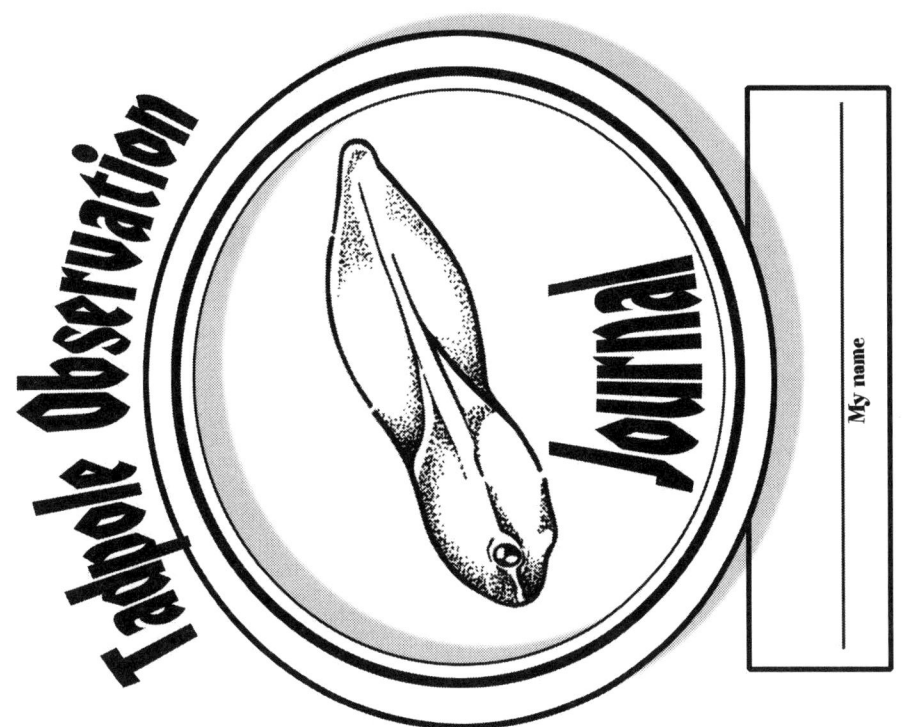

Here are the things I already know about the life cycle of frogs and toads:

2

I learned a lot about how frogs and toads grow. Here are some of the things I didn't know:

Here are some things I would still like to know:

7

A frog or toad begins its life as an _____.

My research says:

Here's how it looks:

3

The tadpole keeps changing as it grows!

Change:

My research says:

Here's how it looks:

6

CRITTERS

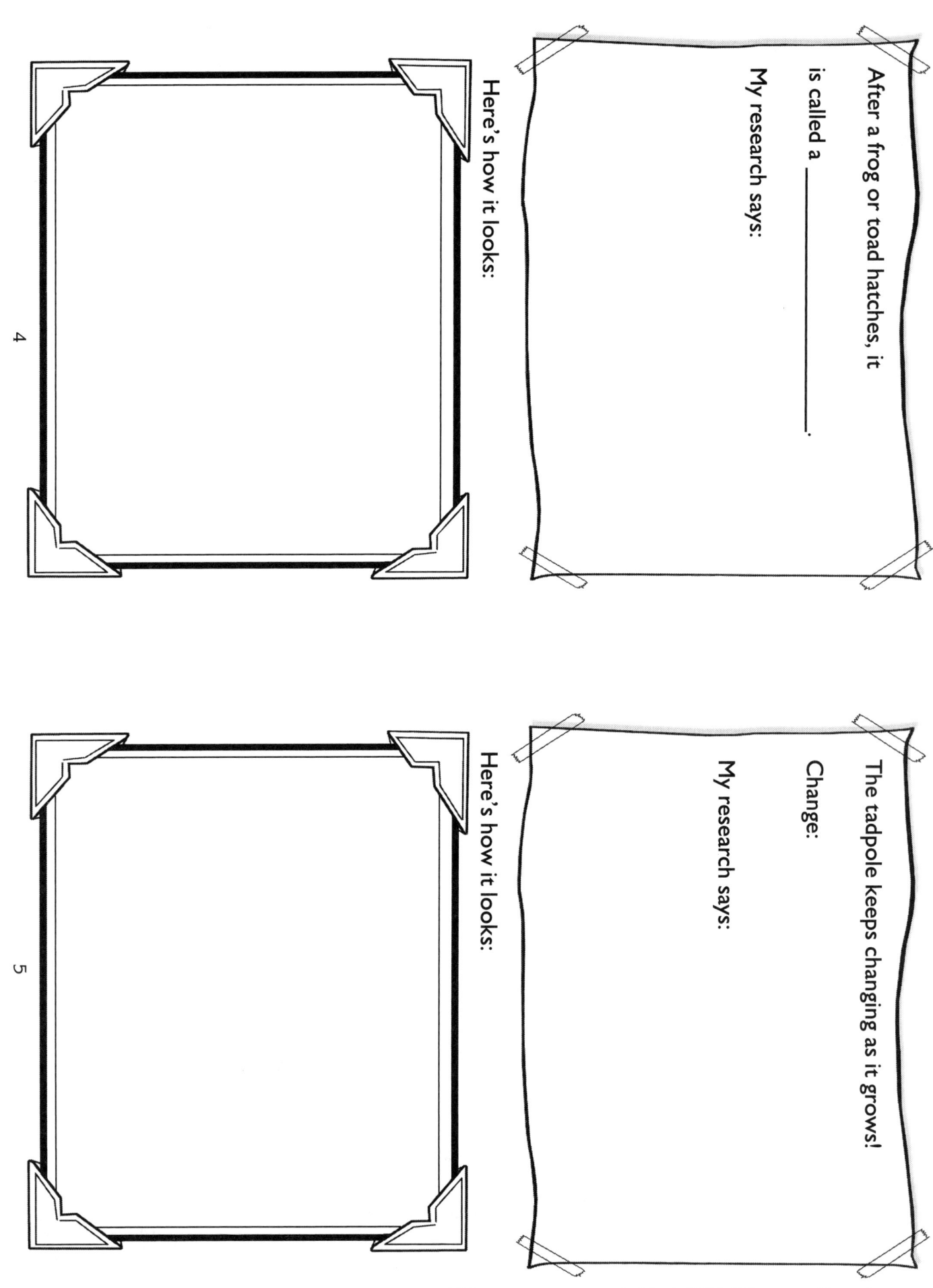

Food Chain

Topic
Food chains and webs

Key Question
Where do plants and animals get the energy they need to survive?

Learning Goals
Students will:
1. learn about food chains and webs by researching animals and where they get their energy,
2. create a food chain or web using the selected animals, and
3. identify the variety of ways transfers of food energy can occur from a source in plants through a series of animals.

Guiding Documents
Project 2061 Benchmarks
- Some source of "energy" is needed for all organisms to stay alive and grow.
- Animals eat plants or other animals for food and may also use plants (or even other animals) for shelter or nesting.
- Insects and various other organisms depend on dead plant and animal material for food.
- Almost all kinds of animals' food can be traced back to plants.
- Two types of organisms may interact with one another in several ways: They may be in a producer/consumer; predator/prey, or parasite/host relationship. Or one organism may scavenge or decompose another. Relationships may be competitive or mutually beneficial. Some species have become so adapted to each other that neither could survive without the other.
- All organisms, including the human species, are part of and depend on two main interconnected global food webs. One includes microscopic ocean plants, the animals that feed on them, and finally the animals that feed on those animals. The other web includes land plants, the animals that feed on them, and so forth. The cycles continue indefinitely because organisms decompose after death to return food materials to the environment.
- Thinking about things as systems means looking for how every part relates to others. The output from one part of a system (which can include material, energy, or information) can become the input to other parts. Such feedback can serve to control what goes on in the system as a whole.
- Locate information in reference books, back issues of newspapers and magazines, compact disks, and computer databases.

NRC Standards
- All animals depend on plants. Some animals eat plants for food. Other animals eat animals that eat the plants.
- Organisms have basic needs. For example, animals need air, water, and food; plants require air, water, nutrients, and light. Organisms can survive only in environments, and distinct environments support the life of different types of organisms.
- An organism's patterns of behavior are related to the nature of that organism's environment, including the kinds and numbers of other organisms present, the availability of food and resources, and the physical characteristics of the environment. When the environment changes, some plants and animals survive and reproduce, and others die or move to new locations.

Science
Environmental science
   interdependence
   food chains and webs

Integrated Processes
Observing
Comparing and contrasting
Collecting and recording data
Organizing
Drawing conclusions

Materials
Paper plates
Scissors
Glue
Brown paper lunch sacks
Set of animal picture cards (included)

CRITTERS © 2004 AIMS Education Foundation

Background Information

A food chain represents the transfer of energy (originating with the sun) from the producer source to a consumer or a series of consumers. For example, a green plant, a leaf-eating insect, and an insect-eating bird would form a simple food chain.

The food web is divided into two broad categories: the grazing web and the detrital web. The grazing web begins with green plants while the detrital web begins with organic debris. Both webs are made up of individual food chains and represent a series of nutritional levels. Green plants, primary producers of food, belong to the first nutritional level, and plant-eating animals belong to the second level. Predators that feed on the plant-eating animals form the third level, and predators that feed on predators belong to the fourth. As the levels rise, the predators become fewer, larger, and fiercer. Seldom are there more than four or five links or levels in a food chain.

Management
1. Students should work together in small groups on this activity.
2. This activity involves research, and will need to take place over a period of a week or more. Students will need access to a variety of research sources such as the Internet, reference books, magazines like National Geographic or Ranger Rick, National Wildlife publications, compact discs, or computer databases.

Procedure
1. Introduce the idea of the energy needed to live and grow by discussing the importance of food for our own survival. Extend the discussion to include all animals.
2. Initiate a "grab-bag research" activity. Use small pictures of animals (included). Place six to eight picture cards in a paper lunch sack and have each group draw out one or two animals for research.
3. Provide groups with the time and materials to find out what kinds of food each animal eats and whether it is eaten by another animal.
4. Direct the groups to use the information gathered to construct food chains using paper plates as the sun and illustrated and titled paper strips to make links showing the path from the original energy source (the sun), to a food source, to a consumer. An example would be: sun —> grain —> field mouse —> owl. One link in each food chain must include an animal drawn from the grab bag.

Connecting Learning
1. What patterns emerged from your research about food chains?
2. Were you able to discover a food chain that did not originate with a plant source?
3. What kinds of animals eat only plants? [herbivores] …only animals? [carnivores] …both plants and animals? [omnivores]
4. Are there any plants that serve as consumers (eat something else, such as animals), or are plants always producers (eaten by something else)? [Some plants are consumers. Decomposers get their energy by breaking down dead plant and animal matter.]
5. How many different food chains can be constructed using the same consumer?
6. How long can a food chain be? [Food chains are rarely longer than five links.] Give an example of a three-link, four-link, and five-link chain.
7. What are the characteristics of the consumers at the higher levels of the food chain? How do these compare to consumers at the lower levels?
8. Organize several food chains to create a food web. How are food chains and webs similar and how are they different?
9. What are you wondering now?

Extension

Construct a giant food web that covers a bulletin board or small wall in your classroom. Write a story that explains how all the parts are interdependent.

Curriculum Correlation

Kalman, Bobbie, and Jacqueline Languille. *What Are Food Chains and Webs?* Crabtree Publishing Company. New York. 1998.

Lauber, Patricia. *Who Eats What?* HarperCollins. New York. 1995.

McKinney, Barbara Shaw. *Pass the Energy, Please!* Dawn Publications. Nevada City, CA. 2000.

Reif, Patricia et. al. *The Magic School Bus Gets Eaten: A Book About Food Chains.* Scholastic, Inc. New York. 1996.

Riley, Peter. *Food Chains.* Franklin Watts. New York. 1999.

Sabin, Francine. *Ecosystems and Food Chains.* Troll Associates. Mahwah, NJ. 1986.

FOOD Chain

1. Color and cut out the sun. Mount it on a paper plate.
2. Color and cut out the food chains.
3. Glue the links together.
4. Hook the links to the sun with paper clips.

Illustrate the links in your food chain. Cut each strip on the solid lines and glue to make a food chain.

glue

glue

glue

glue

glue

Color the links in your food chain. Cut each strip on the solid lines and glue to make a circle. Make a food chain.

glue

glue

glue

glue

glue

glue

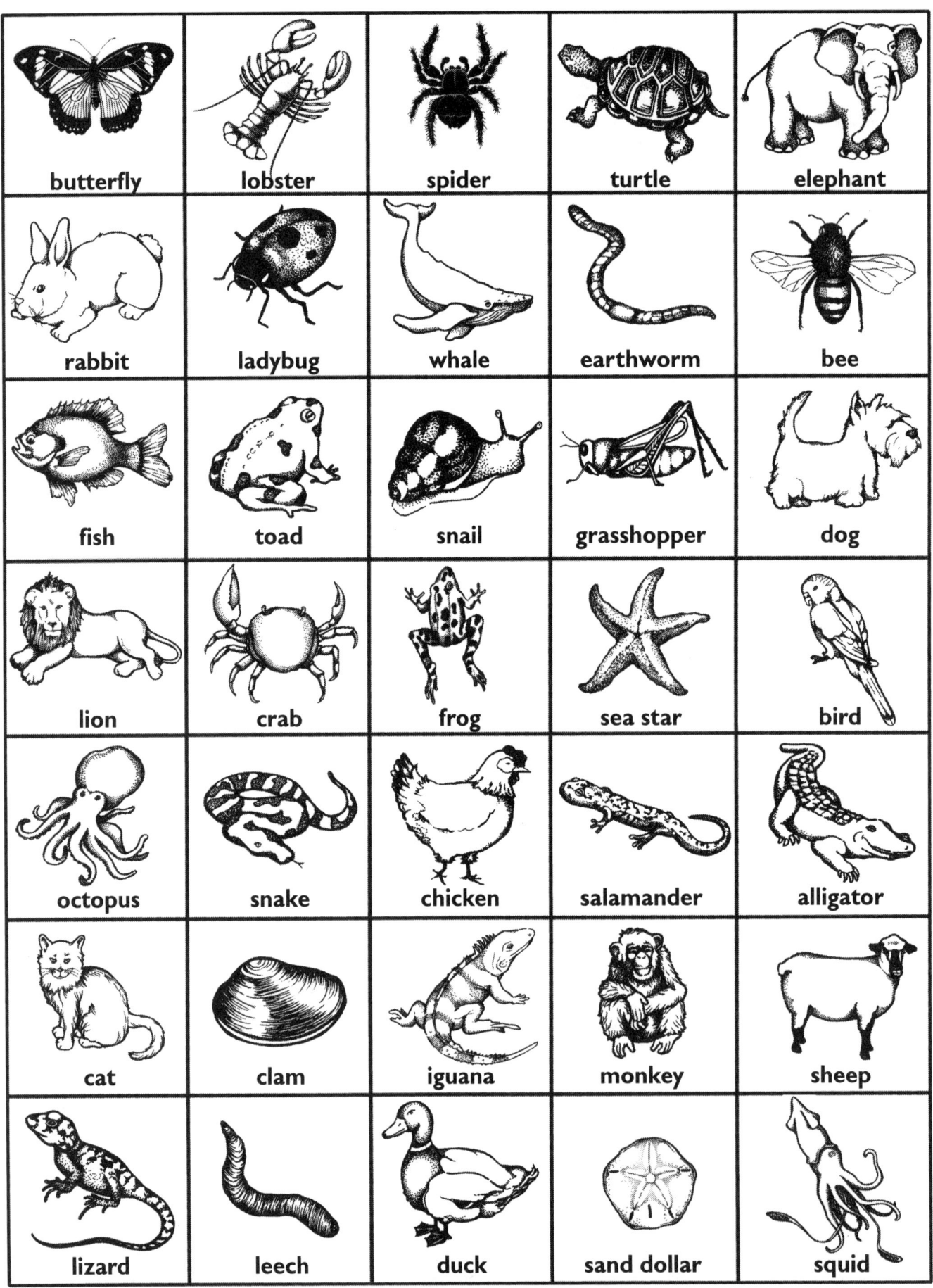

Chain Games

Topic
Food chains

Key Question
What is the primary source of matter and energy entering most food chains?

Learning Goal
Students will learn that plants are the main source of matter and energy entering most food chains.

Guiding Documents
Project 2061 Benchmarks
- *Almost all kinds of animals' food can be traced back to plants.*
- *Food provides the fuel and the building material for all organisms. Plants use the energy from light to make sugars from carbon dioxide and water. This food can be used immediately or stored for later use. Organisms that eat plants break down the plant structures to produce the materials and energy they need to survive. Then they are consumed by other organisms.*
- *Some source of "energy" is needed for all organisms to stay alive and grow.*

NRC Standards
- *All animals depend on plants. Some animals eat plants for food. Other animals eat animals that eat the plants.*
- *Populations of organisms can be categorized by the function they serve in an ecosystem. Plants and some microorganisms are producers—they make their own food. All animals, including humans, are consumers, which obtain food by eating other organisms. Decomposers, primarily bacteria and fungi, are consumers that use waste materials and dead organisms for food. Food webs identify the relationships among producers, consumers, and decomposers in an ecosystem.*
- *For ecosystems, the major source of energy is sunlight. Energy entering ecosystems as sunlight is converted by producers into stored chemical energy through photosynthesis. It then passes from organism to organism in food webs.*

Science
Environmental science
  food chains

Integrated Processes
Observing
Comparing and contrasting
Drawing conclusions
Relating

Materials
Box or wastepaper can filled with scratch paper
Labels (see *Management 3*)
Chain Games Cards
Food Chain Key
Scissors

Background Information
Food chains exist in all habitats and can be used to demonstrate the complexity and energy flow in an ecosystem. Producers capture the sun's energy to make their own food in plant form, while consumers rely on eating those plants or other consumers to get their energy. When an animal eats a plant, it only receives 10% of the energy that the plant got from the sun. Likewise, when an animal eats another animal, it only receives 10% of the energy that animal got from the plants or other things it ate. This 90% energy loss at each level of a food chain is the reason there are so many low-level (primary) consumers and so few top-level consumers.

Management
1. This activity is divided into two parts. In the first part, students will play a game in which they toss a ball of wadded scratch paper to see the loss of energy along the food chain. In the second part, students will play a card game in which they build food chains.
2. Copy the cards on card stock. You may wish to laminate the cards for durability.
3. Make labels on paper or 5" x 7" cards that say *Sun, Grass, Deer, Wolf, Cheese, Mouse,* and *Cat.*
4. The *Chain Games* cards can be linked together in several different ways. For example, a hawk might eat a rabbit, a snake, or a smaller bird. Use the *Food Chain Key* to help identify the things that each animal eats.
5. The concentration card game can be played by two to four students. Adjust the number of card sets needed by the number of groups you have.

CRITTERS

Procedure

Part One

1. Ask the students if they are feeling energetic. Have them name some of the things they have already done today, and ask if doing these things took any energy. See if they can identify a source from which they get all the energy they use. If necessary, guide the discussion to identify food as the energy source, and point out that it takes energy just to do the things we need to stay alive, such as breathing and enabling the other systems of our bodies to function.
2. Ask if they know where the energy in food comes from, making note of some of the possibilities suggested.
3. Select four students to stand side by side in a line in front of the class. Hand a ball made from four pieces of wadded scratch paper to the first student. Ask the students to pass the ball along the line with each person removing a sheet of the paper. The last person should hold on to the single sheet that is left.
4. Explain that the ball represents energy. Distribute labels to identify the first person as the sun, the second as grass, the third as a deer, and the fourth as a wolf. Return the energy ball to the sun and have the students pass it along again as you explain that (1) energy comes from the sun, (2) the grass uses that energy to make and store food in its cells, (3) the deer eats the grass, getting some of the stored energy along with the food matter, and (4) the wolf eats the deer, getting some of the energy that was stored in the deer's cells. Emphasize that the living parts of this arrangement—the plants and the animals—form what is called a food chain.
5. Keep the sun and the grass in place, have students return the wadded paper to make a new energy ball, and ask the other two students to return to their seats. Pick three more students, one to represent cheese (clarifying that cheese is made from milk that comes from a cow), one to represent a mouse, and one to represent a cat. Ask the class to arrange the five students in the order of a food chain. When the order is agreed upon, direct the sun to pass the energy ball along the chain again. As the energy ball moves from link to link, have the students remove paper from the energy ball, and ask them to explain how the energy is being passed along from organism to organism.
6. Ask the class to identify what is the same about the two food chains [sun, grass] and what is different [the animals].
7. Tell the students that the entire class will form a circle around the wastepaper can that contains scratch paper. Inform them that the scratch paper represents energy from the sun. Explain that the student who begins the toss will select four or five pieces of scratch paper and wad them up into a ball that he or she tosses to a classmate. When the classmate catches the ball, the classmate must call out what link he or she represents in the food chain. (The first link needs to be a producer such as grass or leaves.) Then that child removes a piece of the paper and tosses the ball to another student. The student who catches it must call out the name of a consumer and remove a piece of the paper. Students will continue tossing and removing the paper until the chain can go no further (usually three to five links). When the chain is finished, tell the students that they will begin again. Invite a student who has not caught the paper ball to go to the wastepaper can and form a new energy ball to begin the process again.
8. After all students have had the opportunity to catch the energy ball and call out a producer or consumer name, inform the class that they will play a card game to look further into food chains.

Part Two

1. Distribute one set of *Chain Games* cards to each group and have students cut out the cards.
2. Have students shuffle the cards and lay them face down on a desk or table in ordered rows and columns.
3. Explain the rules of the game to students and allow them to play several rounds.

Rules

1. This game is played like the commercial "Memory" or "Concentration" games, but instead of looking for matching cards, players are looking for food chains.
2. Each player's turn begins by turning any two cards face up. If either of these cards begins or continues an existing food chain, the player takes the card(s). For example, any time a player turns over a sun, he/she will take that card because the sun is at the beginning of every food chain.
3. If only one of the two cards can be used by a player, the unusable card is turned face down (in the same location) and the player's turn is over. If a player can use both of the turned over cards, he/she continues to turn cards over, one at a time, until a card that he/she cannot use is turned over. At that point, his/her turn is over, the unusable card is placed face down (in the same location) and the next player's turn begins.
4. All food chains must begin with the sun followed by a producer. The subsequent orders of the animals in the food chains can vary, as indicated by the *Food Chain Key*.

5. Players may have multiple food chains going at the same time. They must always keep all food chain cards they have collected face up and arranged in the correct order. This will allow them, and all of the other players, to quickly determine whether or not the card(s) they turn over can be taken.
6. The game ends when there are no more food chains that can be created using the remaining cards. The player with the most cards at the end of the game is the winner.

Connecting Learning
1. Could plants ever be anywhere in a food chain except at the beginning? [No.] Explain. [They can only get their energy directly from the sun, since they make their own food.]
2. Could animals ever be at the beginning of a food chain? Why or why not? [No, because they can't make their own food, so they have to get their energy by eating a plant or another animal.]
3. Which plants have you noticed at the beginning of several different food chains? Why do you think this is so?
4. Which plants are at the beginning of the food chains that include you?
5. What are you wondering now?

Curriculum Correlation
Kalman, Bobbie, and Jacqueline Languille. *What Are Food Chains and Webs?* Crabtree Publishing Company. New York. 1998.

Lauber, Patricia. *Who Eats What?* HarperCollins. New York. 1995.

McKinney, Barbara Shaw. *Pass the Energy, Please!* Dawn Publications. Nevada City, CA. 2000.

Reif, Patricia et. al. *The Magic School Bus Gets Eaten: A Book About Food Chains.* Scholastic, Inc. New York. 1996.

Riley, Peter. *Food Chains.* Franklin Watts. New York. 1999.

Sabin, Francine. *Ecosystems and Food Chains.* Troll Associates. Mahwah, NJ. 1986.

Sun	Sun	Sun
Sun	Sun	Sun
Grain	Grain	Grain
Leaves	Grass	Grass

Owl 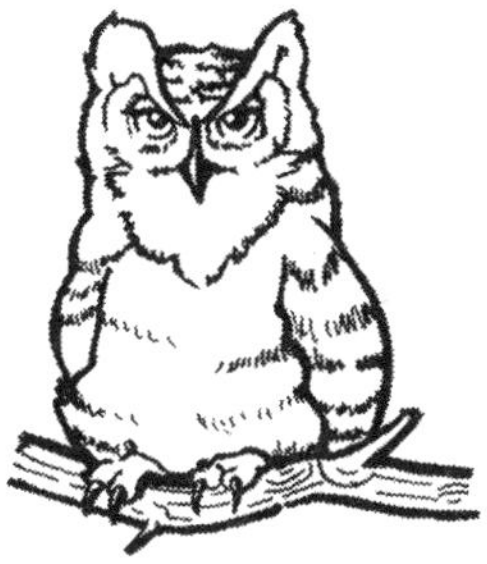	Hawk 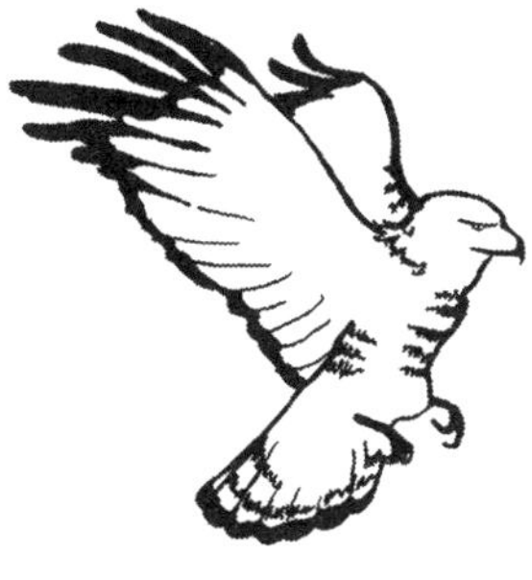	Cat 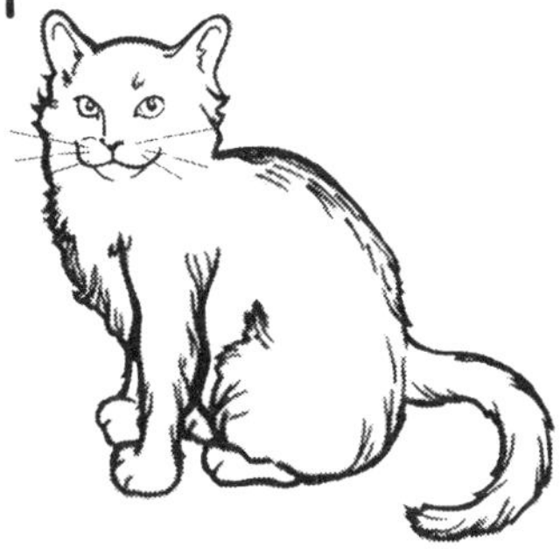
Toad 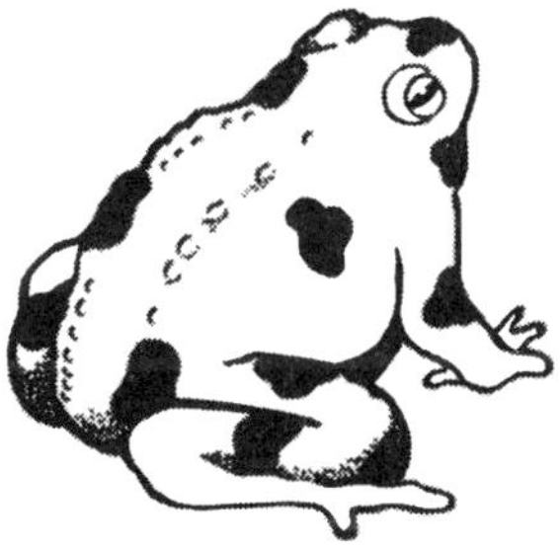	Snake 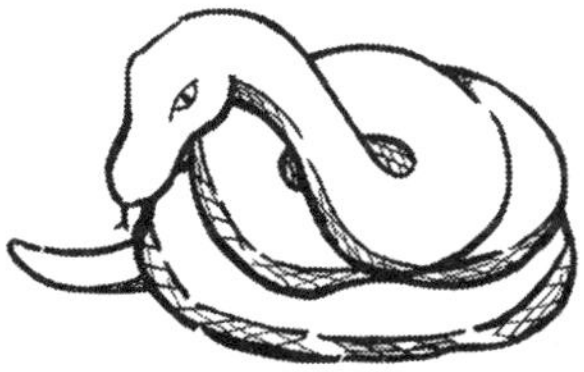	Cow 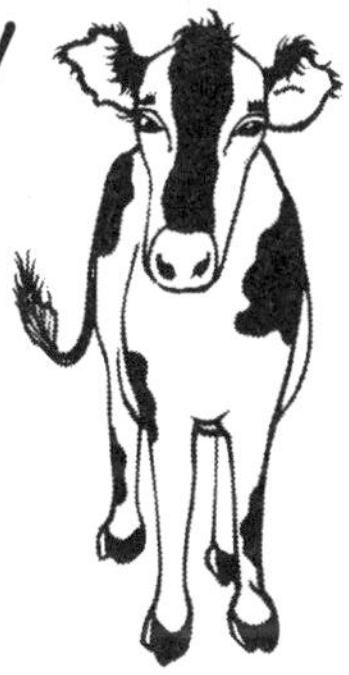
Deer 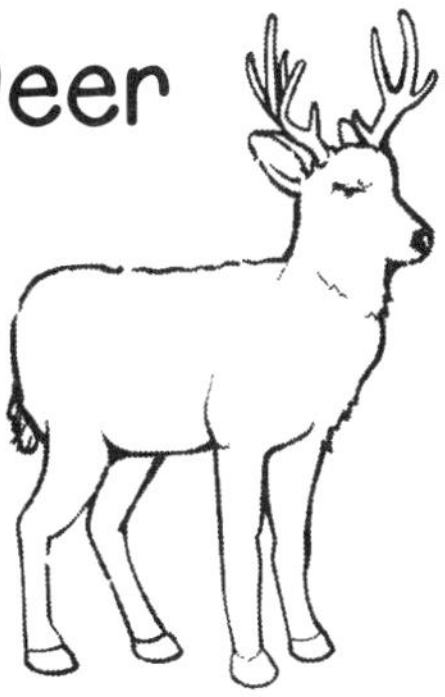	Leaf Beetle 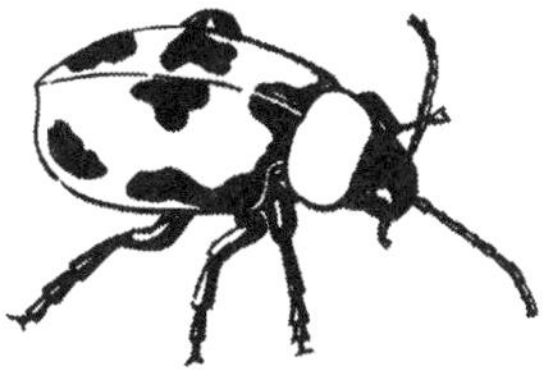	Rabbit 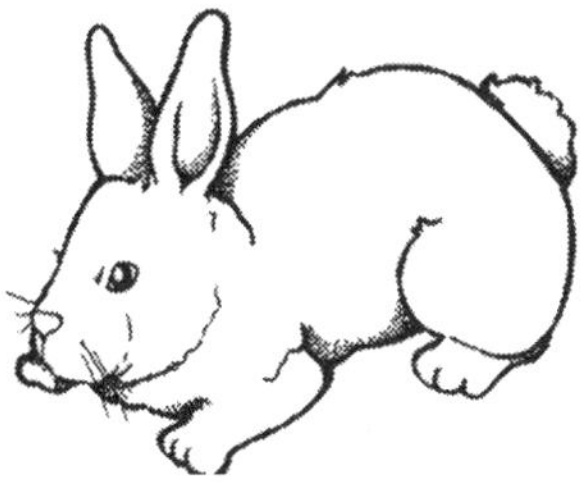
Man 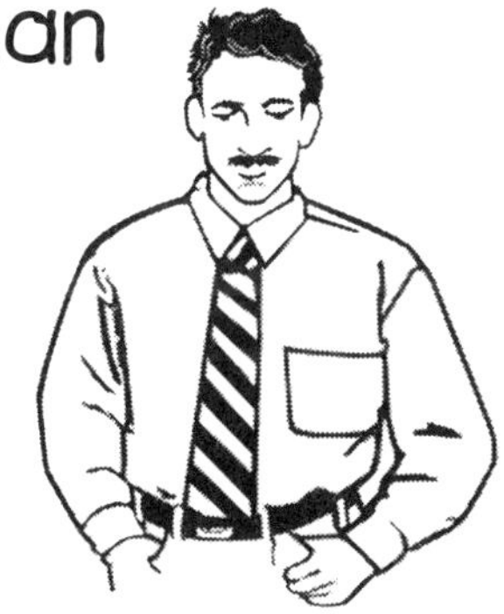	Mouse 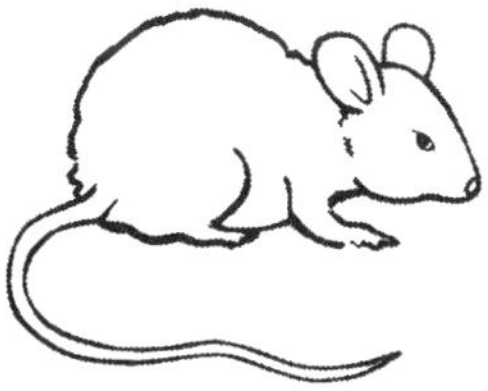	Sparrow 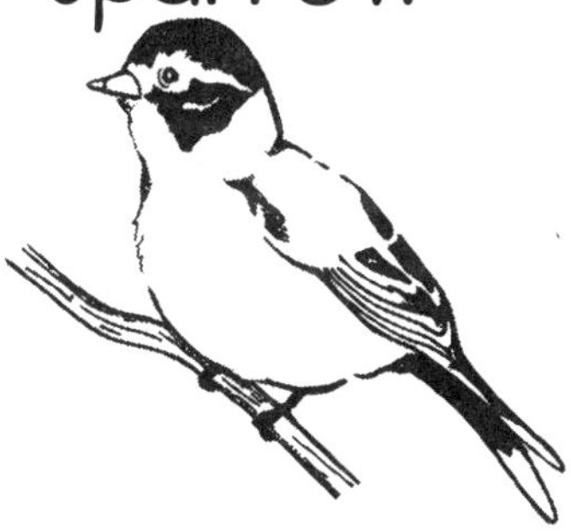

Food Chain Key

Owl	**What it eats:** Frog, Sparrow, Snake, Mouse	Hawk	**What it eats:** Snake, Sparrow, Mouse, Frog, Rabbit	Cat	**What it eats:** Frog, Snake, Sparrow, Rabbit, Mouse
Toad	**What it eats:** Leaf beetle	Snake	**What it eats:** Frog, Mouse	Cow	**What it eats:** Grain, Grass
Deer	**What it eats:** Grain, Leaves, Grass	Leaf Beetle	**What it eats:** Leaves	Rabbit	**What it eats:** Grain, Grass, Leaves
Man	**What it eats:** Cow, Rabbit, Deer, Grain	Mouse	**What it eats:** Grain	Sparrow	**What it eats:** Grain, Leaf beetle

Topic
Food chains

Key Question
How is energy passed along a food chain from link to link?

Learning Goals
Students will:
1. learn about the predator/prey relationship in a food chain, and
2. play a game of tag to experience this relationship.

Guiding Documents
Project 2061 Benchmarks
- *Animals eat plants or other animals for food and may also use plants (or even other animals) for shelter or nesting.*
- *Almost all kinds of animals' food can be traced back to plants.*
- *Some source of "energy" is needed for all organisms to stay alive and grow.*
- *Two types of organisms may interact with one another in several ways: They may be in a producer/consumer; predator/prey, or parasite/host relationship. Or one organism may scavenge or decompose another. Relationships may be competitive or mutually beneficial. Some species have become so adapted to each other that neither could survive without the other.*
- *All organisms, including the human species, are part of and depend on two main interconnected global food webs. One includes microscopic ocean plants, the animals that feed on them, and finally the animals that feed on those animals. The other web includes land plants, the animals that feed on them, and so forth. The cycles continue indefinitely because organisms decompose after death to return food materials to the environment.*
- *Thinking about things as systems means looking for how every part relates to others. The output from one part of a system (which can include material, energy, or information) can become the input to other parts. Such feedback can serve to control what goes on in the system as a whole.*

NRC Standards
- *All animals depend on plants. Some animals eat plants for food. Other animals eat animals that eat the plants.*
- *Organisms have basic needs. For example, animals need air, water, and food; plants require air, water, nutrients, and light. Organisms can survive only in environments, and distinct environments support the life of different types of organisms.*
- *An organism's patterns of behavior are related to the nature of that organism's environment, including the kinds and numbers of other organisms present, the availability of food and resources, and the physical characteristics of the environment. When the environment changes, some plants and animals survive and reproduce, and others die or move to new locations.*

Science
Life science
    food chains
        predator/prey relationship

Integrated Processes
Observing
Comparing and contrasting
Collecting and recording data
Identifying and manipulating variables
Generalizing
Analyzing

Materials
Brown, yellow and red yarn (see *Management 3*)
Large bag of plain, popped popcorn
Plastic sandwich bags, one per student
Cheese popcorn, optional

Background Information
All foods contain chemical energy. A food chain shows the transfer of energy through the chain. Energy is released from the sun and converted by green plants (producers in the food chain) that use light to make food through photosynthesis. Primary consumers are dependent on green plants, and thus the sun, for food energy. Higher level consumers are dependent on the animals that eat plants or other animals, thus the energy is passed from link to link in the food chain. All links are ultimately dependent on the sun for their food energy. Some energy is lost between each link in a food chain. Because of the energy loss, each higher level has fewer living things than the level below it. This means that food chains rarely exceed four links.

A pyramid of energy, or biomass pyramid, illustrates the energy transfer between predators and prey. Animals at the top of the pyramid are fewer in number and need to eat many smaller animals to get enough energy to survive. The primary consumers that feed on green plants are much more numerous. In a well-balanced ecosystem, the producers and consumers at each level have numbers that are large enough to ensure their survival without depleting their food supply, thus the pyramid effect with many producers and primary consumers and few of the highest level consumers.

Management
1. Find an area with well-defined boundaries for this outdoor activity.
2. Stress safety and demonstrate the proper way to tag. Make sure students understand the rules before going outside.
3. You will need yarn in brown, red, and yellow cut into pieces about 40 cm long. Cut enough brown for half the class, enough yellow for one-third of the class, and enough red for one-third of the class.
4. Use the activity sheet after the final round of play.

Procedure
1. Review food chains and food webs. Discuss predator/prey relationships.
2. Tell students they are going to play a game of tag that will simulate a natural food chain and illustrate a biomass pyramid.
3. Divide the class into three even groups. Each group will be assigned a different color of yarn. Hand out the yarn and have students help each other to tie the yarn around their wrists in a bow that can be easily removed at the end of the game.
4. Explain that the animals the students are simulating are represented by the colors of yarn. Students with brown yarn are grasshoppers, those with yellow yarn are lizards, and those with red yarn are hawks.
5. Discuss the predator/prey relationships in this food chain. Hawks hunt only lizards. Lizards hunt only grasshoppers. Grasshoppers eat only grass (which is represented by the popcorn).
6. Give each student a plastic bag to be used as a stomach and explain how the game will work. Those students who are grasshoppers must gather popcorn from the ground and put in their plastic bags. The students playing lizards will try to tag the grasshoppers. If they are successful, the grasshopper is "dead" and the contents of his/her bag are emptied into the lizard's bag. (The empty bag stays with the grasshopper to be used again in the next round.) The students playing hawks will try to tag the lizards, and get the contents of their bags if successful. Lizards and hawks may not pick up popcorn from the ground.
7. For the animals to survive, they must not be tagged during the game and their stomachs (plastic bags) must be filled by the game's end as follows:
grasshoppers—1/3 full
lizards—2/3 full
hawks—full
8. Go outdoors and select an area to be the ecosystem. For the first round, the area should be small so that the students can experience the effects of crowding on animal populations. Students may not leave the area during the game.
9. Set up two or three safe zones within the selected area. These zones should be large enough for two students at a time. Whenever a new student enters a safe zone, the one who has been there the longest must leave. Animals may not prey upon each other in these zones. Select another area in which the "dead" animals can wait for the next round.
10. Spread out a large bag of popped popcorn over the ecosystem.
11. Signal the primary consumers, the grasshoppers, to begin eating grass (gathering popcorn). After 30 seconds, allow the lizards to enter the area. After 30 more seconds, allow the hawks to enter the ecosystem. Allow the students to play for several minutes or until no prey are left. At the end of play, all remaining animals must have the right amount of food in their plastic bags or they too are dead. Note the length of time the game lasted.
12. After this first round, ask why the game only lasted a few minutes. Discuss crowding and the number of predators vs. the number of prey. Write down the number and kinds of animals that are still alive.
13. Play the second round in the same area as the first one with the following changes: have half the students play grasshoppers, and divide the other half so that two-thirds of them are lizards and one-third are hawks. Play the game again. Discuss the effects that changing the population numbers had on the time the game lasted.
14. For the third round, leave the animal populations as they were in round two, but greatly enlarge the area in which the game is played. Discuss the effects of the larger area on the time the game lasted.
15. Return to the classroom. Use the activity sheet to illustrate the numbers of predators and prey in an ecosystem and to make a biomass mobile.

Discuss the energy flow from the producers to the higher level consumers. Emphasize that the energy in a food chain originates from the sun. A biomass pyramid could also be made by centering and gluing the pieces, one on top of another.

Connecting Learning

1. Why did the games end? [all the prey was dead] Which games were the shortest? [those with the fewest prey] ...longest? [those with the most prey] Why? [The more prey there are the longer it takes for the predators to eat them all.]
2. What numbers of predators and prey worked the best? [more prey, fewer predators]
3. How does area affect predator/prey relationships?
4. How is this game like a real ecosystem? [It shows one food chain and how the predator/prey relationship works.] How is it different? [There would be more variables in a real ecosystem.]
5. Where does grass get energy? [the sun]
6. Where does the grasshopper get energy? [grass] ...the lizard [grasshopper] ...the hawk? [lizard]
7. How is the biomass mobile related to the predator/prey game? Why are green plants so important in a food chain?
8. What are you wondering now?

Extensions

1. Mix some cheese popcorn in with the regular popcorn to represent a pesticide. Do not tell students what it represents until the end of the game. When the round is over inform them that any animal with three or more pieces of cheese popcorn in its stomach is dead due to toxic poisoning.
2. Create a variety of food chains using other animals.
3. Play the game introducing predator/prey behaviors such as camouflage, hunting techniques, decoying, running speed, freezing, and playing dead.

Curriculum Correlation

Literature
Kalman, Bobbie, and Jacqueline Languille. *What Are Food Chains and Webs?* Crabtree Publishing Company. New York. 1998.

Lauber, Patricia. *Who Eats What?* HarperCollins. New York. 1995.

McKinney, Barbara Shaw. *Pass the Energy, Please!* Dawn Publications. Nevada City, CA. 2000.

Reif, Patricia et. al. *The Magic School Bus Gets Eaten: A Book About Food Chains.* Scholastic, Inc. New York. 1996.

Riley, Peter. *Food Chains.* Franklin Watts. New York. 1999.

Sabin, Francine. *Ecosystems and Food Chains.* Troll Associates. Mahwah, NJ. 1986.

Geography
Research various geographical areas and list several food chains found there that are different than those found in your area.

Art
Design a poster illustrating food chains and food webs.

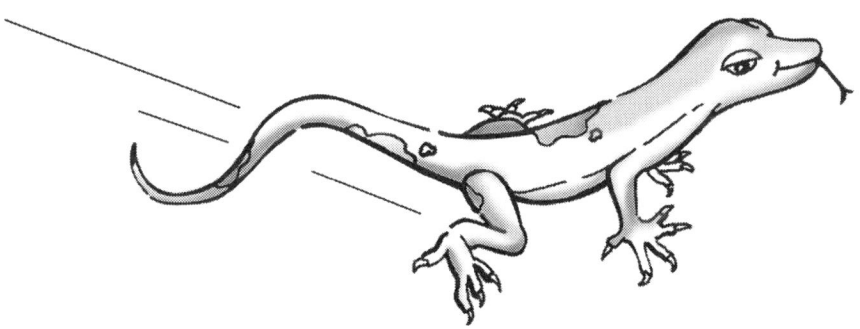

Catch Me if You Can

Square	Length	Width	Area
A		x	=
B		x	=
C		x	=
D		x	=

Find the length, width, and area of each square. Record this information in the table.

Imagine that the squares are part of an ecosystem that includes grass, grashoppers, lizards, and hawks. Think about the numbers of living things in a balanced ecosystem and color the squares according to this key.

lizards = yellow   grasshoppers = brown   grass = green   hawks = red

A

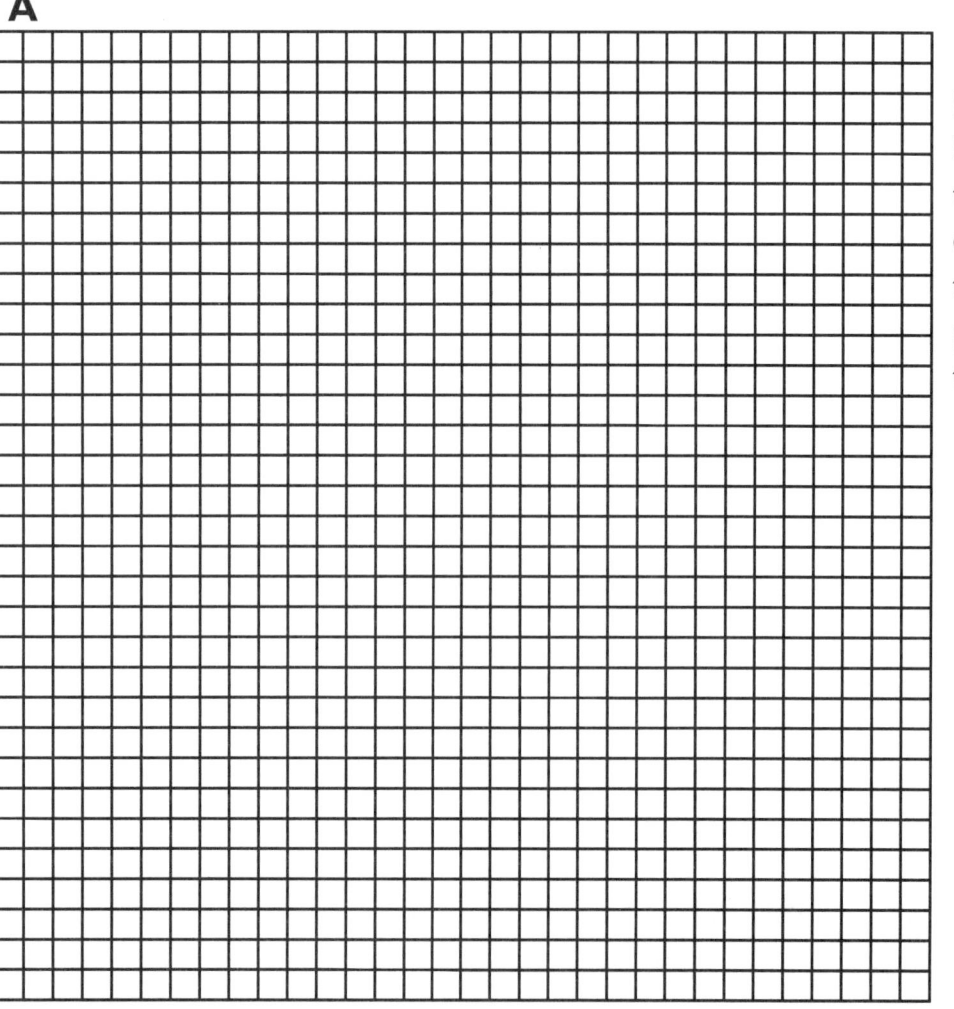

Make a mobile by cutting out the squares and connecting them with a piece of thread.

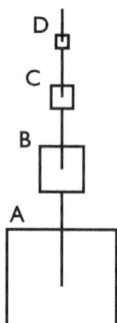

B

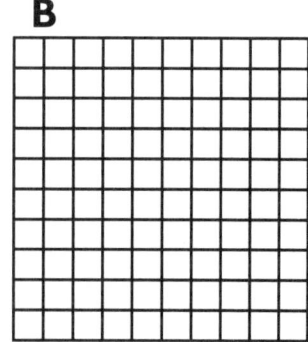

C

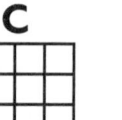

D

CRITTERS

Census Takers

Topic
Population sampling

Key Question
How can you find the population of a large number of animals?

Learning Goals
Students will:
1. explore the concept of sampling,
2. estimate and then take samples of a critter population,
3. compare the actual population total to the estimates, and
4. make a line graph of the results.

Guiding Document
NCTM Standards 2000*
- *Collect data using observations, surveys, and experiments*
- *Represent data using tables and graphs such as line plots, bar graphs, and line graphs*
- *Propose and justify conclusions and predictions that are based on data and design studies to further investigate the conclusions or predictions*
- *Design investigations to address a question and consider how data-collection methods affect the nature of the data set*

Math
Sampling
Estimation
Graphing
   bar graph
Averages

Science
Life science
   animal populations

Integrated Processes
Observing
Comparing and contrasting
Collecting and recording data

Materials
Activity sheets
Scissors

Background Information
Knowing the population of an organism is important in studying it. Many times it is very difficult or impossible to count a large population of organisms. To get a good idea of the total population of an organism, scientists determine the area the organism occupies. They then count the number of organisms in random, small sections of the larger area. The numbers of organisms in the samples are then averaged to find the average number of organisms living in a unit area. This unit area varies with the size and range of the organism being studied. For large organisms with a big range, the unit might be square kilometers. For smaller organisms, it might be square meters, and for even smaller organisms, it could be square centimeters or millimeters. When the average number of organisms per unit area is found, it is used to estimate the total population by multiplying the number of organisms in the unit area by the total area the organism occupies. This gives scientists a fairly accurate population estimate for a given organism.

Management
1. Students can work in small groups or individually.
2. Students should only count whole or nearly whole critters within the cut-out square when they take their samples.
3. After taking their samples and estimating, students may count the actual number of critters on the sheet or be given the correct number (450).

Procedure
1. Introduce the concept of sampling. This might be done by asking the class how many students are in their school and brainstorming ways that they could find out. Discuss ways to get an answer without counting every student. Students could then do a population study of the school by finding the average number of students per class by counting a few classes and averaging (sampling) and multiplying that number times the number of classes in the school. Compare the results from this method to the actual count from the office.
2. Distribute the activity sheet with the critters drawn on it and have students predict the total number of critters.

CRITTERS   197   © 2004 AIMS Education Foundation

3. Hand out the second activity sheet and have each student record his or her estimate. Have students cut out their sampling squares and use them to take five samples of the population. Be sure the samples are randomly taken. Students may drop the cut-out square on the critter sheet or close their eyes and place it on the sheet. Have students record the number of critters counted in each sample and calculate the average number of critters in a square unit. They multiply the average number of critters for the sample square unit, by 56, since the critter population inhabits an area of 56 square units (7 x 8).
4. Compare the population estimates to the actual count (450 critters) and find the difference.
5. Record each group member's population estimates on the board and make a bar graph.
6. Using a different color crayon or pencil, color a bar on the graph to represent the actual number of critters. Have the students compare their estimates to the line representing the actual number.
7. Using the group estimates, find the class average of the population estimates and compare it to the actual count.

Connecting Learning
1. Why is it important to select a random sample?
2. How did the population estimates and the actual population differ?
3. Do you think this is an accurate way to determine the population of a large area?
4. Why would this be a particularly good method for finding whale populations?
5. How is a real plot study different from our classroom activity?
6. How does the class average compare to the actual population?
7. What are you wondering now?

Extensions
1. Sprinkle sand on a sheet of graph paper and take several samples and find the approximate number of grains on the whole sheet.
2. Have students do a plot study of a local area.

Curriculum Correlation
Geography
Discuss local plants and animals that could be used to do a census.

Science
Discuss how plot studies may differ at different times of the day and during different seasons.

* Reprinted with permission from *Principles and Standards for School Mathematics*, 2000 by the National Council of Teachers of Mathematics. All rights reserved.

Census Takers

Procedure:

- Carefully cut out the square *Critter Counter* at the bottom of this page. Be as exact as you can.
- Look at the page of critters. Estimate the total number of critters on the page. Record your estimate in the table below.
- Randomly drop your cut-out square on the page of critters, trace around it, and count the number you see within the square. Record the number in the table below. Do this five times. Add your five samples together to get a sample total.

Estimate	Population samples					Sample total
	A	B	C	D	E	

- Divide your sample total by five to get the average. Multiply the average by the number of square units on the page of critters to determine the population estimate.

Sample total	÷ 5 =	Average	x	Number of square units	=	Population sample estimate
	÷ 5 =		x		=	

- Record the actual population and find the difference between your estimate and the actual. If your estimate is higher than the actual population, subtract the actual from the estimate.

Actual population	−	Population estimate	=	Difference
	−		=	

CRITTERS

Graph the estimates of each member of your group. Write each person's name on a line at the bottom of the graph.
Use one column for the estimate and the second column for the actual.

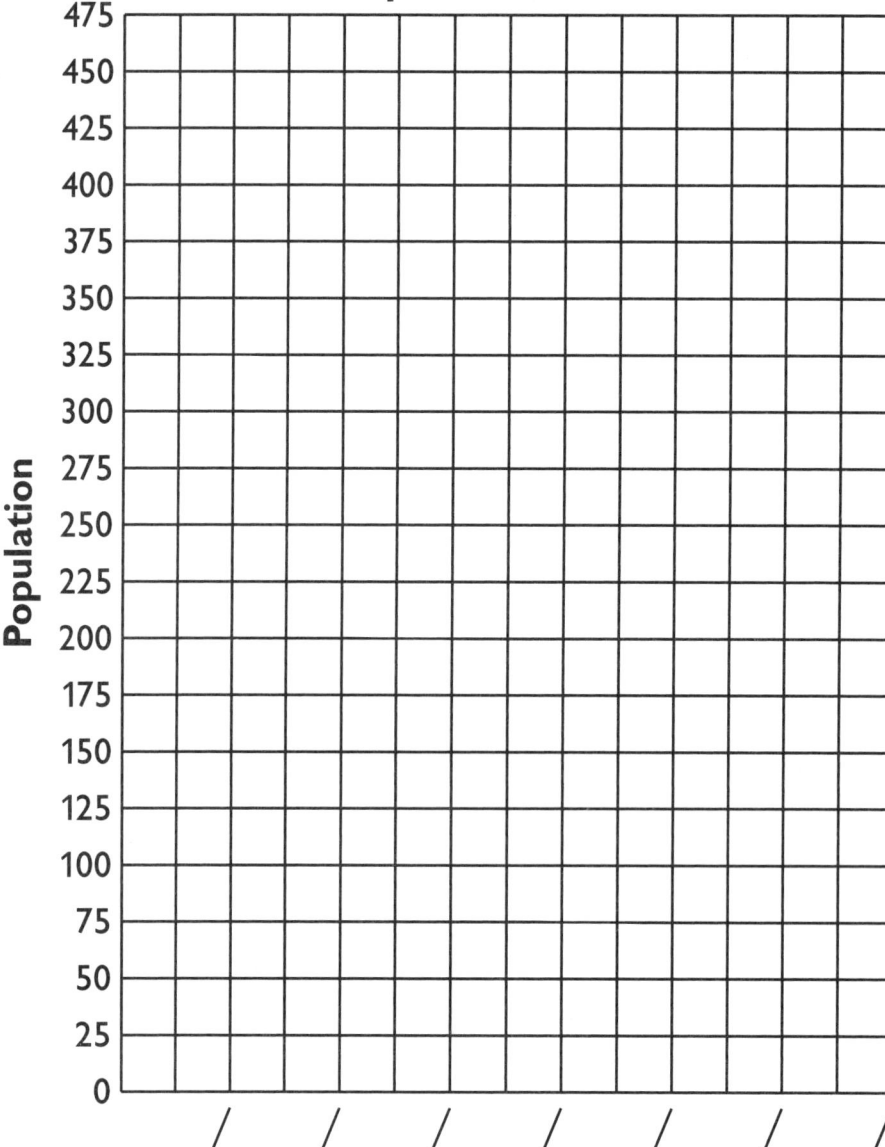

Population Estimates

Census Takers

CRITTERS 200 © 2004 AIMS Education Foundation

Once you have collected your data and completed your graph, answer the following questions.

1. How close was your first population estimate to the actual total?

2. How close was your population sample estimate to the actual total? Why?

3. Why was it important that you randomly placed the square on the paper for each of your samples?

4. Was the random sampling method an effective way to estimate the population? Why or why not?

5. What was the difference between your population estimate and the actual population?

6. What did the graph show you about other people's population estimates?

7. Do you think our method would be an effective way to determine the population of an animal in a large area? Why or why not?

Topic
Biomes

Key Questions
1. What are the different biomes on Earth?
2. What kinds of plants and animals live in each biome?

Learning Goals
Students will:
1. learn about eight of the Earth's biomes,
2. discover how the different biomes meet the needs of the plants and animals that live in them, and
3. make a diorama of a biome showing its plant and animal life.

Guiding Documents
Project 2061 Benchmarks
- *Living things are found almost everywhere in the world. There are somewhat different kinds in different places.*
- *For any particular environment, some kinds of plants and animals survive well, some survive less well, and some cannot survive at all.*
- *Locate information in reference books, back issues of newspapers and magazines, compact disks, and computer databases.*

NRC Standard
- *Organisms have basic needs. For example, animals need air, water, and food; plants require air, water, nutrients, and light. Organisms can survive only in environments, and distinct environments support the life of different types of organisms.*

Science
Environmental science
   ecology
      biomes

Integrated Processes
Observing
Collecting and recording data
Researching

Materials
Biome mini-books
Student page
Materials for dioramas (see *Management 2*)
Research materials (see *Management 4*)

Background Information
Biomes are large areas that have similar geography, climate, plants, and animals. The amount of rainfall and the temperature are two important ways in which biomes are divided. There are land (terrestrial) and water (aquatic) biomes. There is not agreement in academic circles on how many biomes exist. Some people divide the Earth into as few as five biomes, while others are more specific, listing more than 15. We have chosen to focus on eight of the broadest and most widely accepted biomes, recognizing that others may have different categories.

Management
1. Students need to be divided into eight groups for this activity. Each group will be responsible for one biome.
2. Provide whatever materials are necessary for students to construct their biome dioramas. It is suggested that students make large dioramas, using something the size of a paper box rather than the more traditional shoebox.
3. This activity will need to take place over several days or weeks so that students have sufficient time to conduct research and create their dioramas.
4. In addition to the mini-books provided on each biome, you will want to have resources your students can use to look up plant and animal life in their biomes. See the *Internet Connections* and *Curriculum Correlation* sections for suggested websites and books.

Procedure
1. Ask the *Key Question* and state the *Learning Goals*.
2. Divide students into their working groups and assign one biome to each group. Distribute the appropriate mini-book to each group.
3. Give each child a copy of the student page to guide the research.
4. Provide sufficient time for students to research their biome and the plants and animals that live there.
5. When each group has completed its diorama, conduct a time of class sharing where each group tells the rest of the class about their biome.

Connecting Learning
1. Where is your biome located?
2. What are the characteristics of your biome (rainfall, climate, etc.)?
3. What kinds of plants live in your biome? How does the biome meet their needs?
4. What kinds of animals live in your biome? How does the biome meet their needs?
5. Could a _____ (name an animal from the rainforest) survive in the tundra? Why or why not?
6. Could a _____ (name an animal from the ocean) survive in the freshwater biome? Why or why not?
7. What are you wondering now?

Extension
Have students find out what biome(s) exist in your area. Find additional native plants and animals that live there.

Internet Connections
Missouri Botanical Garden
http://mbgnet.mobot.org/
Great kid-friendly site uses the same biome categories as this activity. Has lots of great information, maps, photographs, and more.

The World's Biomes
http://www.ucmp.berkeley.edu/glossary/gloss5/biome/
This site divides the biomes into five broad categories: aquatic, deserts, forests, grasslands, and tundra. Within each broad category, it describes the many sub-categories in depth.

Curriculum Correlation
Literature
Johansson, Philip. *The Dry Desert: A Web of Life.* Enslow Publishers, Inc. Berkeley Heights, NJ. 2004.

Johansson, Philip. *The Forested Taiga: A Web of Life.* Enslow Publishers, Inc. Berkeley Heights, NJ. 2004.

Johansson, Philip. *The Frozen Tundra: A Web of Life.* Enslow Publishers, Inc. Berkeley Heights, NJ. 2004.

Johansson, Philip. *The Temperate Forest: A Web of Life.* Enslow Publishers, Inc. Berkeley Heights, NJ. 2004.

Johansson, Philip. *The Tropical Rain Forest: A Web of Life.* Enslow Publishers, Inc. Berkeley Heights, NJ. 2004.

Johansson, Philip. *The Wide Open Grasslands: A Web of Life.* Enslow Publishers, Inc. Berkeley Heights, NJ. 2004.

Kalman, Bobbie. *What is a Biome?* Crabtree Publishing. New York. 1998.

What is a biome?
A biome is an area on Earth that has similar geography, climate, plants, and animals. There are land biomes and water biomes.

How many biomes are there?
Not everyone agrees about how many biomes there are. Some people break the Earth into five biomes. Others say there are more than 15. For this activity, we will be looking at eight biomes: grasslands, taiga, tundra, tropical rain forests, freshwater, saltwater, deserts, and temperate forests.

My Group's Research

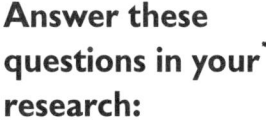

_____
Our Biome

Answer these questions in your research:

1. Where is your biome located? Include countries as well as continents. If your biome falls between certain lines of latitude, list those.

2. Describe the characteristics of your biome. Include average rainfall, summer and winter temperatures, and geography.

3. List at least five plants that live in your biome.

4. Describe how your biome meets the needs of these plants.

5. List at least five animals that live in your biome. Be sure to include as many kinds of animals as you can (birds, mammals, insects, fish, amphibians, reptiles, etc.)

6. Describe how your biome meets the needs of these animals.

Desert biomes receive very little annual rainfall, usually less than 25 cm per year. Most people think of deserts as being hot places, but there are cold deserts too! The Gobi desert in Mongolia is a cold desert. Deserts are found on every continent, even Antarctica!

The plants and animals that live in deserts have special adaptations that allow them to survive with very little water. Many plants in hot deserts have expansive root systems. Others have the capacity to store water. Some reduce the surface area that can lose water to evaporation. Common desert plants include cacti, yucca trees, prickly pears, creosote bushes, and brittle bushes.

Animals that are common in hot deserts include kangaroo rats, desert tortoises, jackrabbits, lizards, snakes, coyotes, and owls. Cold desert animals include llamas, bighorn sheep, pronghorn antelopes, jackals, and lizards.

Fresh water all over the world makes up this biome. It includes rivers, lakes, streams, ponds, swamps, and wetlands.

Algae are plants common to all these bodies of water. Elodea, duckweed, and water lilies are other freshwater plants.

Many animals live in the water as well as along the shoreline. Beavers, all kinds of fish, frogs, turtles, alligators, crayfish, snails, salamanders, mosquitoes, and dragonflies are found in the freshwater biome.

GRASSLANDS

The animals that live in the grass-lands vary widely depending on location. Residents of grasslands include elephants, gazelles, giraffes, zebras, buffalo, lions, hyenas, leopards, gophers, rabbits, antelope, wild horses, prairie dogs, jackrabbits, quail, ring-necked pheasants, meadowlarks, grasshoppers, hawks, owls, snakes, moles, ground squirrels, and mice.

Grasslands cover about one-fourth of the land on Earth. The grasslands are called different things in different places. They include prairies, steppes, plains, pampas, savannas, and veldts. Grasslands can be found in Africa, Australia, North America, South America, and Asia.

These areas get as much as 127 cm of rain a year, or as little as 15 cm, depending on the location. The plant life in this biome is mostly grasses. There are few or no trees or large shrubs.

SALTWATER

In the deeper waters, there are whales, sharks, squid, swordfish, sea turtles, octopuses, sea lions, dolphins, and tuna. The ocean is the only biome that contains no insects, although they may fly over its surface.

The saltwater biome is the largest of all biomes. Approximately seven-tenths of the Earth's surface is water, and most of that is salt water. Any water environment that has at least 3.5% salt is considered salt water. Scientists divide the ocean into four smaller biomes based on the amount of light and the temperature.

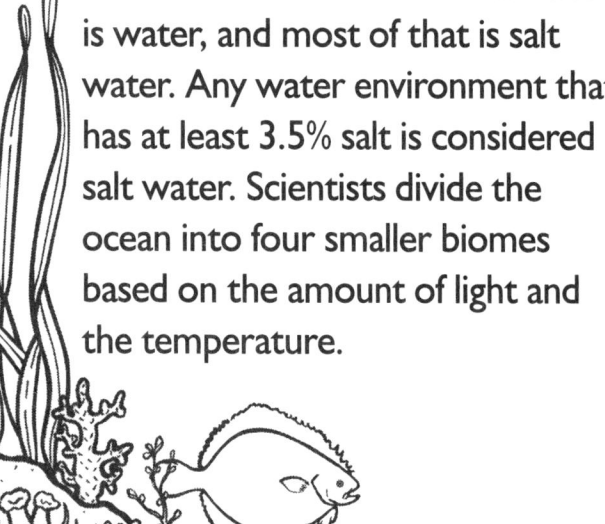

Algae, bacteria, plankton, kelp, and seaweed are found in the ocean biome. In tropical oceans, there are coral reefs. Animals found in the ocean are varied depending upon where they live. Clams, crabs, barnacles, flounders, mussels, oysters, sea cucumbers, sea stars, and stingrays live close to shore.

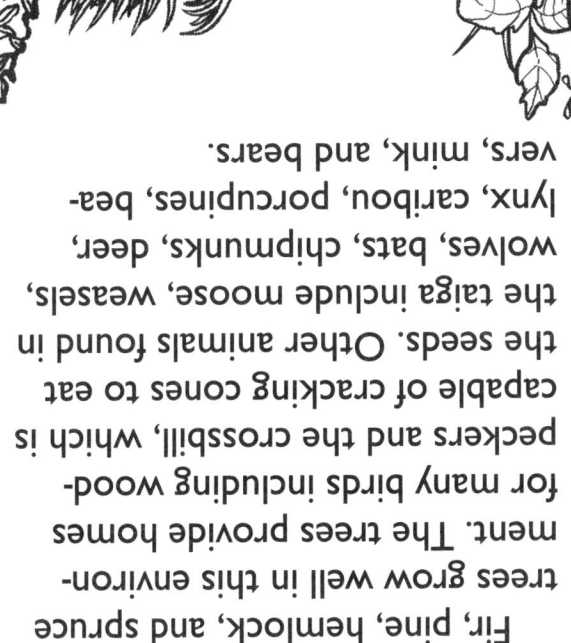

Fir, pine, hemlock, and spruce trees grow well in this environment. The trees provide homes for many birds including woodpeckers and the crossbill, which is capable of cracking cones to eat the seeds. Other animals found in the taiga include moose, weasels, wolves, bats, chipmunks, deer, lynx, caribou, porcupines, beavers, mink, and bears.

TAIGA

The taiga, or boreal forest, is the world's largest land biome. It is mostly located between the latitudes 50° N and 60° N. It stretches across large portions of North America, Europe, and Asia. The average yearly rainfall is between 30 and 85 centimeters.

Winters in the taiga are long, cold, and dry. The days are very short. During the summers, the days are long. It is warm and wet, and the soil is good for lichens and mosses. Insects thrive in the taiga during the summers.

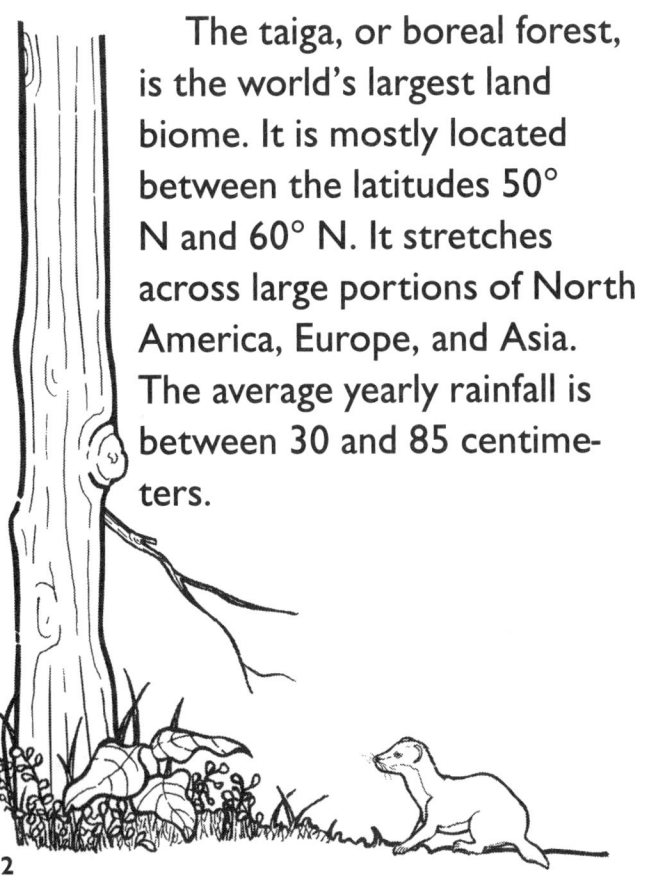

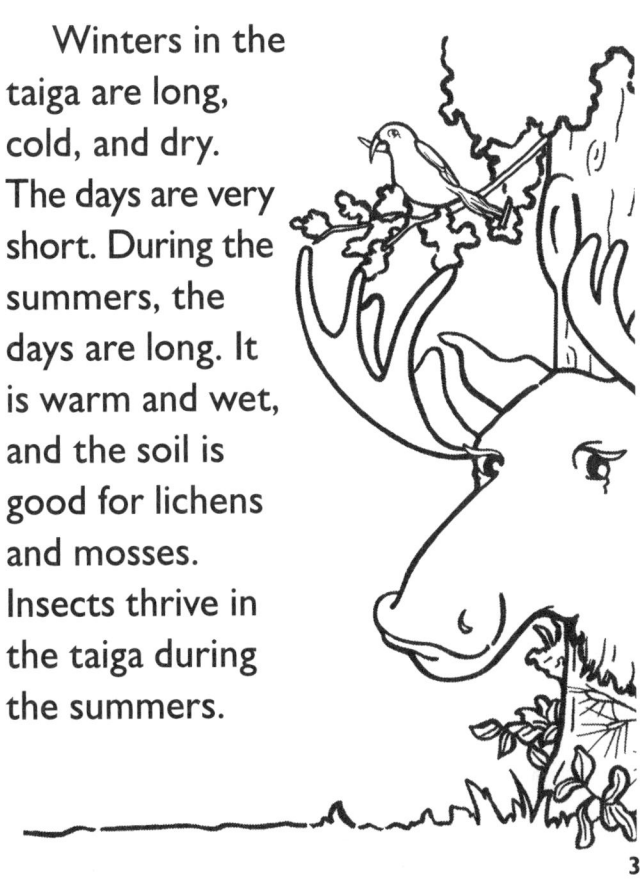

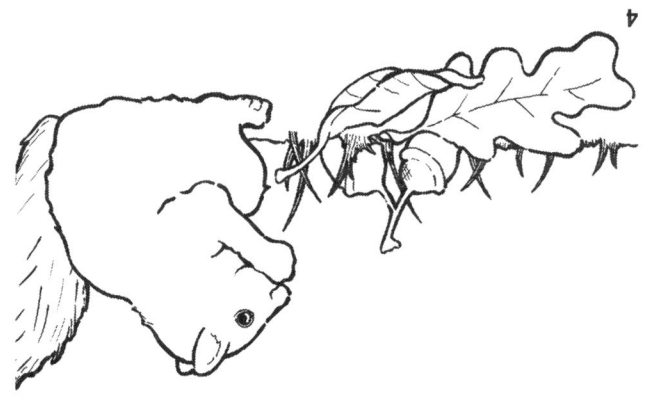

Visitors will find worms, insects, salamanders, and snakes on the ground. Many birds like cardinals, woodpeckers, and turkeys live in temperate forests. Mammals include squirrels, raccoons, rabbits, skunks, opossum, bears, mountain lions, foxes, and deer.

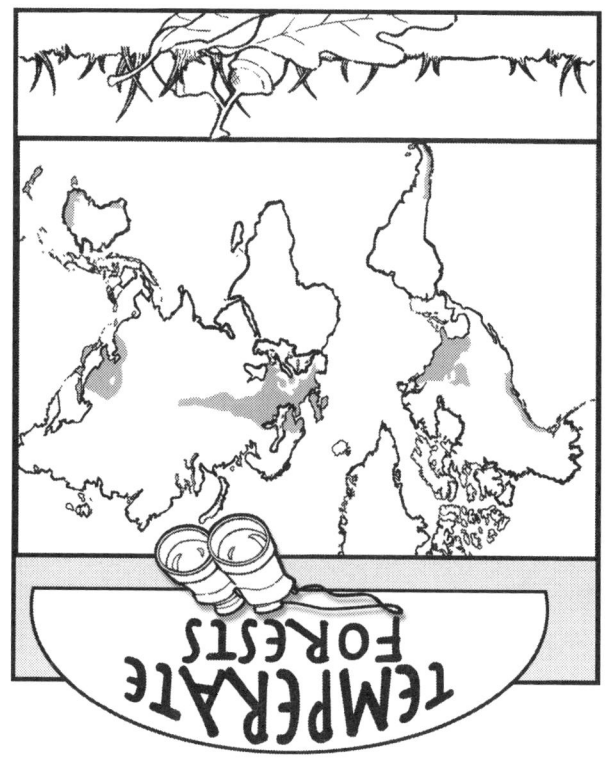

The temperate forest consists of areas that have an annual rainfall between 75 and 150 centimeters. These areas are found primarily in North America, Asia, and Europe. This biome has four distinct seasons, but rarely has a snow cover all winter.

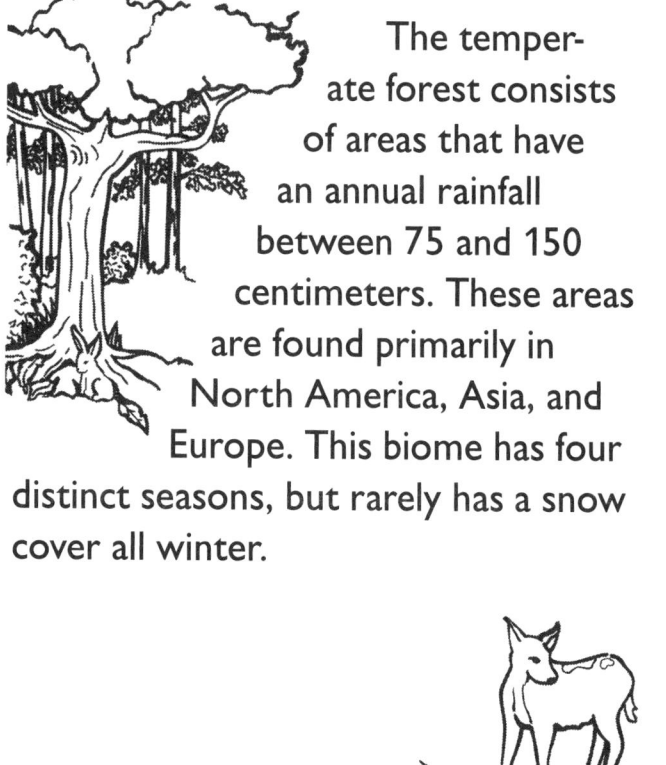

Many trees, including beeches, maples, oaks, elms, willows, and hickories grow in the temperate forest. All of these trees are *deciduous*. This means that they lose their leaves during the fall.

TROPICAL RAIN FOREST

Many of the rainforest's animals also live in the trees. Bats, squirrels, monkeys, tree frogs, and sloths are found here. On the ground there are gorillas, beetles, millipedes, spiders, snakes, and antelope.

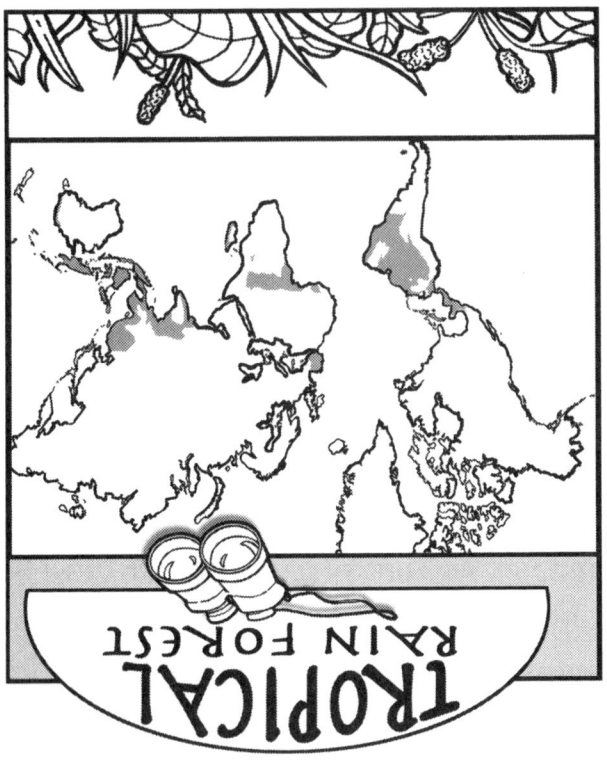

The tropical rain forest is found around the equator (between latitudes 23.5° N and 23.5° S). Temperatures are about 25° C all year. The average yearly rainfall is 200 cm. There are only two seasons in a tropical rain forest—rainy and dry.

Portions of South America, Central America, Central Asia, Australia, and Africa are included in this biome. Because of the humid conditions, this biome has the greatest amount of plant growth, and is extremely diverse. Large trees such as banyans, wild figs, and mahogany form a canopy that prevents much light from reaching the ground. Other plants like bromeliads, orchids, and lianas live in the canopy.

The tundra is the world's coldest biome. The tundra is divided into two categories: arctic and alpine. Arctic tundra is found around the North Pole. Alpine tundra exists at high altitudes (above the tree line) on mountains around the world. The tundra gets very little precipitation (15 to 25 cm per year). Most of its water comes from snow.

The arctic tundra has a layer of ground called permafrost. This layer remains frozen year round. Because of the severe weather conditions, there are limited producers in the tundra. These include lichen, reindeer moss, true mosses, liverworts, tussock grasses, and some low shrubs and sedges. Trees cannot establish root systems through the permafrost.

Although there is a variety of animal life, many animals live in the tundra only during the summer. Caribou, polar bears, foxes, marmots, mountain goats, elk, wolves, lemmings, squirrels, falcons, salmon, trout, snowshoe hares, mice, and moles can be found there. Flies and mosquitoes have adapted their life cycles to the extremes of the tundra weather and live in the tundra year round.

TUNDRA

Who's Home in the Biome?

Topic
Biomes

Key Question
What plants and animals live in each biome?

Learning Goal
Students will review which plants and animals are found in each biome by playing a card game.

Guiding Documents
Project 2061 Benchmarks
- *Living things are found almost everywhere in the world. There are somewhat different kinds in different places.*
- *For any particular environment, some kinds of plants and animals survive well, some survive less well, and some cannot survive at all.*

NRC Standard
- *Organisms have basic needs. For example, animals need air, water, and food; plants require air, water, nutrients, and light. Organisms can survive only in environments, and distinct environments support the life of different types of organisms.*

Science
Environmental science
    ecology
        biomes

Integrated Processes
Observing
Comparing and contrasting
Classifying

Materials
Per group of four:
    biome mini-books (see *Management 2*)
    biome plant/animal record cards
    32 index cards (see *Management 4*)
    colored pencils/markers

Background Information
Biomes are large areas that have similar geography, climate, plants, and animals. The amount of rainfall and the temperature are two important ways in which biomes are divided. There are land (terrestrial) and water (aquatic) biomes. There is not agreement in academic circles on how many biomes exist. Some people divide the Earth into as few as five biomes, while others are more specific, listing more than 15. We have chosen to focus on eight of the broadest and most widely accepted biomes, recognizing that others may have different categories.

Management
1. This activity is meant to be a follow-up to a more in-depth study of biomes. It is assumed that students have completed *Biome Boxes* before they do this activity.
2. Students will need their biome mini-books and research pages from *Biome Boxes*.
3. Cut out the biome plant and animal record cards before starting the game.
4. Each student will need eight blank cards on which to draw and label the chosen plants or animals from his/her biome. Index cards (3" x 5") work well.
5. If students have a hard time drawing the plants and animals on the blank playing cards, they can paste pictures of plants and animals cut from old magazines on the blank cards instead. The playing cards should be labeled with the name of the plant or animal.

Procedure
1. Place students in groups of four.
2. Have each student select one of the eight biomes. Be sure that no two students in the same group choose the same biome.
3. If students select a biome other than the one they researched in *Biome Boxes*, give them the appropriate mini-book.
4. Give students time to read the mini-books and choose eight plants and/or animals that live in that biome.
5. Distribute the biome record cards, blank playing cards, and colored pencils to each group.
6. Have each student record the eight choices on his or her biome record card. Instruct them to draw pictures of the plants and animals chosen on the blank playing cards. Be sure that they label these pictures. If necessary, provide pictures of plants and animals that students can cut and paste.
7. Explain the rules and allow groups to play the game.

Rules

1. Shuffle all the plant and animal playing cards together.
2. Deal three playing cards to each player and place the rest in a pile face down in the center of the group.
3. One player begins by asking any other player for a plant or animal from his/her biome record card. If the other player has the requested card, it must be given to the player who asked for it. The first player then gets to go again.
4. If the other player does not have the requested card, he/she replies with, "Go search." The first player then draws a card from the top of the pile. If the player draws the card requested, he/she gets to go again. If not, the next player takes a turn.
5. Play continues in this fashion until someone collects five out of the eight cards listed on his/her biome record card and calls out "Biome."

Connecting Learning

1. Did anyone have the same plant or animal as another player?
2. How is it possible for a plant or animal to live in more than one biome?
3. What plants and animals are restricted to a single biome? Why?
4. Which biomes have the most limited selection of plants and animals? Why?
5. What are you wondering now?

Extensions

1. Have students create their own games using the biome cards.
2. Play a larger game using cards from all eight biomes.

Internet Connections

Missouri Botanical Garden
http://mbgnet.mobot.org/
Great kid-friendly site uses the same biome categories as this activity. Has lots of good information, maps, photographs, and more.

The World's Biomes
http://www.ucmp.berkeley.edu/glossary/gloss5/biome/
This site divides the biomes into five broad categories: aquatic, deserts, forests, grasslands, and tundra. Within each broad category it describes the many sub-categories in depth.

Curriculum Correlation

Literature
Johansson, Philip. *The Dry Desert: A Web of Life.* Enslow Publishers, Inc. Berkeley Heights, NJ. 2004.

Johansson, Philip. *The Forested Taiga: A Web of Life.* Enslow Publishers, Inc. Berkeley Heights, NJ. 2004.

Johansson, Philip. *The Frozen Tundra: A Web of Life.* Enslow Publishers, Inc. Berkeley Heights, NJ. 2004.

Johansson, Philip. *The Temperate Forest: A Web of Life.* Enslow Publishers, Inc. Berkeley Heights, NJ. 2004.

Johansson, Philip. *The Tropical Rain Forest: A Web of Life.* Enslow Publishers, Inc. Berkeley Heights, NJ. 2004.

Johansson, Philip. *The Wide Open Grasslands: A Web of Life.* Enslow Publishers, Inc. Berkeley Heights, NJ. 2004.

Kalman, Bobbie. *What is a Biome?* Crabtree Publishing. New York. 1998.

Geography
Research what biomes you would travel through if you traveled from the North Pole to the South Pole. Is there a pattern?

Ecologist: _____	Ecologist: _____
Biome: _____	Biome: _____
1. _____	1. _____
2. _____	2. _____
3. _____	3. _____
4. _____	4. _____
5. _____	5. _____
6. _____	6. _____
7. _____	7. _____
8. _____	8. _____

Ecologist: _____	Ecologist: _____
Biome: _____	Biome: _____
1. _____	1. _____
2. _____	2. _____
3. _____	3. _____
4. _____	4. _____
5. _____	5. _____
6. _____	6. _____
7. _____	7. _____
8. _____	8. _____

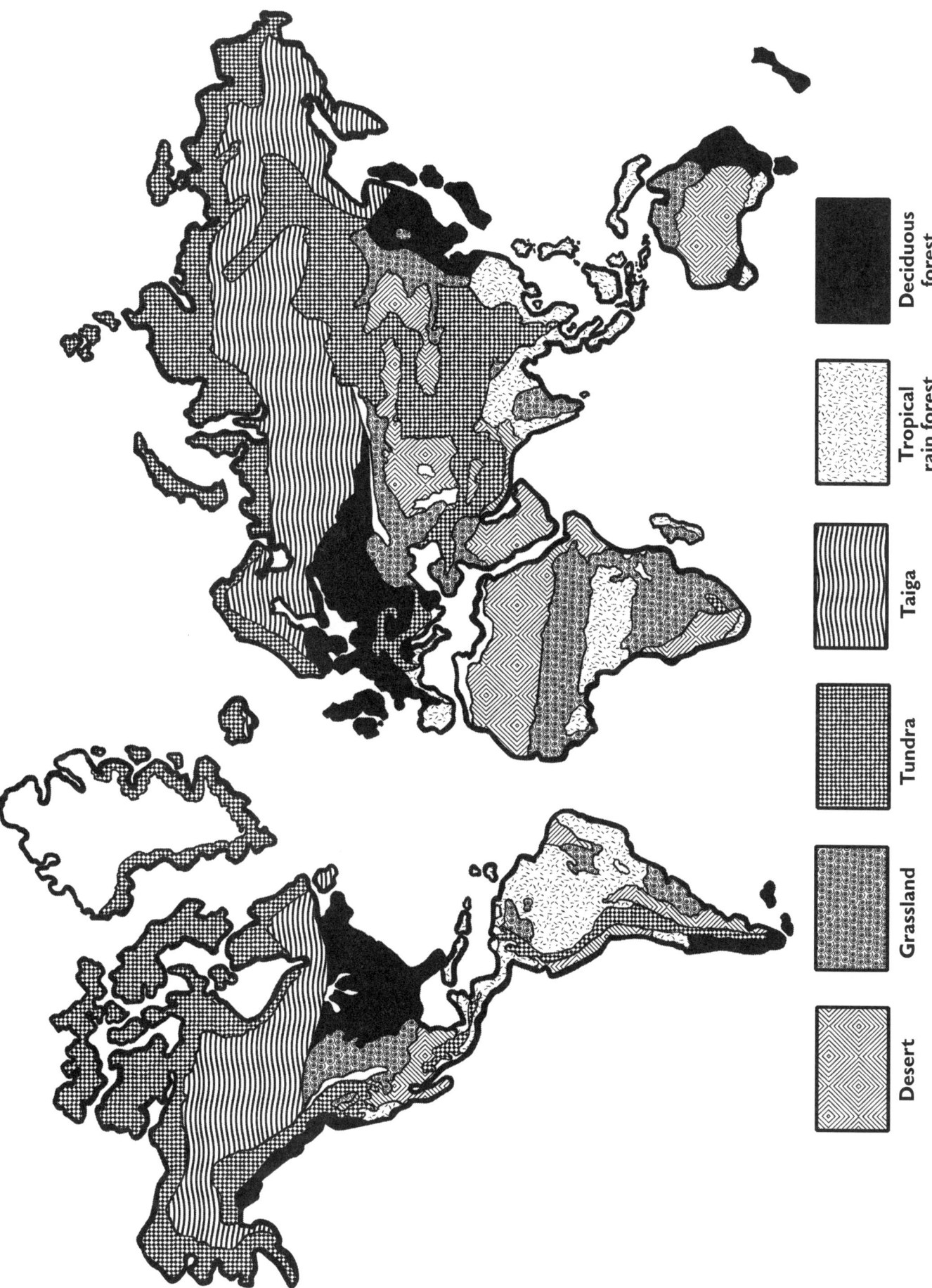

The AIMS Program

AIMS is the acronym for "Activities Integrating Mathematics and Science." Such integration enriches learning and makes it meaningful and holistic. AIMS began as a project of Fresno Pacific University to integrate the study of mathematics and science in grades K-9, but has since expanded to include language arts, social studies, and other disciplines.

AIMS is a continuing program of the non-profit AIMS Education Foundation. It had its inception in a National Science Foundation funded program whose purpose was to explore the effectiveness of integrating mathematics and science. The project directors in cooperation with 80 elementary classroom teachers devoted two years to a thorough field-testing of the results and implications of integration.

The approach met with such positive results that the decision was made to launch a program to create instructional materials incorporating this concept. Despite the fact that thoughtful educators have long recommended an integrative approach, very little appropriate material was available in 1981 when the project began. A series of writing projects ensued, and today the AIMS Education Foundation is committed to continuing the creation of new integrated activities on a permanent basis.

The AIMS program is funded through the sale of books, products, and staff development workshops, and through proceeds from the Foundation's endowment. All net income from programs and products flows into a trust fund administered by the AIMS Education Foundation. Use of these funds is restricted to support of research, development, and publication of new materials. Writers donate all their rights to the Foundation to support its on-going program. No royalties are paid to the writers.

The rationale for integration lies in the fact that science, mathematics, language arts, social studies, etc., are integrally interwoven in the real world, from which it follows that they should be similarly treated in the classroom where students are being prepared to live in that world. Teachers who use the AIMS program give enthusiastic endorsement to the effectiveness of this approach.

Science encompasses the art of questioning, investigating, hypothesizing, discovering, and communicating. Mathematics is a language that provides clarity, objectivity, and understanding. The language arts provide us with powerful tools of communication. Many of the major contemporary societal issues stem from advancements in science and must be studied in the context of the social sciences. Therefore, it is timely that all of us take seriously a more holistic method of educating our students. This goal motivates all who are associated with the AIMS Program. We invite you to join us in this effort.

Meaningful integration of knowledge is a major recommendation coming from the nation's professional science and mathematics associations. The American Association for the Advancement of Science in *Science for All Americans* strongly recommends the integration of mathematics, science, and technology. The National Council of Teachers of Mathematics places strong emphasis on applications of mathematics found in science investigations. AIMS is fully aligned with these recommendations.

Extensive field testing of AIMS investigations confirms these beneficial results:

1. Mathematics becomes more meaningful, hence more useful, when it is applied to situations that interest students.
2. The extent to which science is studied and understood is increased when mathematics and science are integrated.
3. There is improved quality of learning and retention, supporting the thesis that learning which is meaningful and relevant is more effective.
4. Motivation and involvement are increased dramatically as students investigate real-world situations and participate actively in the process.

We invite you to become part of this classroom teacher movement by using an integrated approach to learning and sharing any suggestions you may have. The AIMS Program welcomes you!

AIMS Education Foundation Programs

Practical proven strategies to improve student achievement

When you host an AIMS workshop for elementary and middle school educators, you will know your teachers are receiving effective, usable training that they can apply in their classrooms immediately.

Designed for teachers—AIMS Workshops:
- Correlate to your state standards;
- Address key topic areas, including math content, science content, problem solving, and process skills;
- Teach you how to use AIMS' effective hands-on approach;
- Provide practice of activity-based teaching;
- Address classroom management issues, higher-order thinking skills, and materials;
- Give you AIMS resources; and
- Offer college (graduate-level) credits for many courses.

Aligned to district and administrator needs—AIMS workshops offer:
- Flexible scheduling and grade span options;
- Custom (one-, two-, or three-day) workshops to meet specific schedule, topic, and grade-span needs;
- Pre-packaged one-day workshops on most major topics—only $3900 for up to 30 participants (includes all materials and expenses);
- Pre-packaged week-long workshops (four- or five-day formats) for in-depth math and science training—only $12,300 for up to 30 participants (includes all materials and expenses);
- Sustained staff development, by scheduling workshops throughout the school year and including follow-up and assessment;
- Eligibility for funding under the Eisenhower Act and No Child Left Behind; and
- Affordable professional development—save when you schedule consecutive-day workshops.

University Credit—Correspondence Courses

AIMS offers correspondence courses through a partnership with Fresno Pacific University.
- Convenient distance-learning courses—you study at your own pace and schedule. No computer or Internet access required!

The tuition for each three-semester unit, graduate-level course is $264 plus a materials fee.

The AIMS Instructional Leadership Program

This is an AIMS staff-development program seeking to prepare facilitators for leadership roles in science/math education in their home districts or regions. Upon successful completion of the program, trained facilitators become members of the AIMS Instructional Leadership Network, qualified to conduct AIMS workshops, teach AIMS in-service courses for college credit, and serve as AIMS consultants. Intensive training is provided in mathematics, science, process and thinking skills, workshop management, and other relevant topics.

Introducing AIMS Science Core Curriculum

Developed in alignment with your state standards, AIMS' Science Core Curriculum gives students the opportunity to build content knowledge, thinking skills, and fundamental science processes.
- Each grade specific module has been developed to extend the AIMS approach to full-year science programs.
- Each standards-based module includes math, reading, hands-on investigations, and assessments.

Like all AIMS resources, these core modules are able to serve students at all stages of readiness, making these a great value across the grades served in your school.

For current information regarding the programs described above, please complete the following:

Information Request

Please send current information on the items checked:

___ *Basic Information Packet* on AIMS materials ___ Hosting information for AIMS workshops
___ *AIMS Instructional Leadership Program* ___ AIMS Science Core Curriculum

Name _____ Phone _____

Address_____
         Street   City   State   Zip

AIMS Magazine
YOUR K-9 MATH AND SCIENCE CLASSROOM ACTIVITIES RESOURCE

The AIMS Magazine is your source for standards-based, hands-on math and science investigations. Each issue is filled with teacher-friendly, ready-to-use activities that engage students in meaningful learning.

- *Four issues each year (fall, winter, spring, and summer).*

Current issue is shipped with all past issues within that volume.

| 1820 | Volume XX | 2005-2006 | $19.95 |
| 1821 | Volume XXI | 2006-2007 | $19.95 |

Two-Volume Combination

| M20507 | Volumes XX & XXI | 2005-2007 | $34.95 |

Back Volumes Available
Complete volumes available for purchase:

1802	Volume II	1987-1988	$19.95
1804	Volume IV	1989-1990	$19.95
1805	Volume V	1990-1991	$19.95
1807	Volume VII	1992-1993	$19.95
1808	Volume VIII	1993-1994	$19.95
1809	Volume IX	1994-1995	$19.95
1810	Volume X	1995-1996	$19.95
1811	Volume XI	1996-1997	$19.95
1812	Volume XII	1997-1998	$19.95
1813	Volume XIII	1998-1999	$19.95
1814	Volume XIV	1999-2000	$19.95
1815	Volume XV	2000-2001	$19.95
1816	Volume XVI	2001-2002	$19.95
1817	Volume XVII	2002-2003	$19.95
1818	Volume XVIII	2003-2004	$19.95
1819	Volume XIX	2004-2005	$35.00

Call today to order back volumes: 1.888.733.2467.

Call 1.888.733.2467 or go to www.aimsedu.org

Subscribe to the AIMS Magazine

$19.95 a year!

AIMS Magazine is published four times a year.

Subscriptions ordered at any time will receive all the issues for that year.

AIMS Online – www.aimsedu.org

For the latest on AIMS publications, tips, information, and promotional offers, check out AIMS on the web at www.aimsedu.org. Explore our activities, database, discover featured activities, and get information on our college courses and workshops, too.

AIMS News

While visiting the AIMS website, sign up for AIMS News, our FREE e-mail newsletter. Published semi-monthly, AIMS News brings you food for thought and subscriber-only savings and specials. Each issue delivers:

- Thought-provoking articles on curriculum and pedagogy;
- Information about our newest books and products; and
- Sample activities.

Sign up today!

AIMS Program Publications

Actions with Fractions, 4-9
Awesome Addition and Super Subtraction, 2-3
Bats Incredible! 2-4
Brick Layers II, 4-9
Chemistry Matters, 4-7
Counting on Coins, K-2
Cycles of Knowing and Growing, 1-3
Crazy about Cotton, 3-7
Critters, 2-5
Electrical Connections, 4-9
Exploring Environments, K-6
Fabulous Fractions, 3-6
Fall into Math and Science, K-1
Field Detectives, 3-6
Finding Your Bearings, 4-9
Floaters and Sinkers, 5-9
From Head to Toe, 5-9
Fun with Foods, 5-9
Glide into Winter with Math and Science, K-1
Gravity Rules! 5-12
Hardhatting in a Geo-World, 3-5
It's About Time, K-2
It Must Be A Bird, Pre-K-2
Jaw Breakers and Heart Thumpers, 3-5
Looking at Geometry, 6-9
Looking at Lines, 6-9
Machine Shop, 5-9
Magnificent Microworld Adventures, 5-9
Marvelous Multiplication and Dazzling Division, 4-5
Math + Science, A Solution, 5-9
Mostly Magnets, 2-8
Movie Math Mania, 6-9
Multiplication the Algebra Way, 4-8
Off the Wall Science, 3-9
Out of This World, 4-8
Paper Square Geometry:
   The Mathematics of Origami, 5-12
Puzzle Play, 4-8
Pieces and Patterns, 5-9
Popping With Power, 3-5
Positive vs. Negative, 6-9
Primarily Bears, K-6
Primarily Earth, K-3
Primarily Physics, K-3
Primarily Plants, K-3

Problem Solving: Just for the Fun of It! 4-9
Problem Solving: Just for the Fun of It! Book Two, 4-9
Proportional Reasoning, 6-9
Ray's Reflections, 4-8
Sense-Able Science, K-1
Soap Films and Bubbles, 4-9
Solve It! K-1: Problem-Solving Strategies, K-1
Solve It! 2nd: Problem-Solving Strategies, 2
Solve It! 3rd: Problem-Solving Strategies, 3
Spatial Visualization, 4-9
Spills and Ripples, 5-12
Spring into Math and Science, K-1
The Amazing Circle, 4-9
The Budding Botanist, 3-6
The Sky's the Limit, 5-9
Through the Eyes of the Explorers, 5-9
Under Construction, K-2
Water Precious Water, 2-6
Weather Sense: Temperature, Air Pressure, and Wind, 4-5
Weather Sense: Moisture, 4-5
Winter Wonders, K-2

Spanish Supplements*
Fall Into Math and Science, K-1
Glide Into Winter with Math and Science, K-1
Mostly Magnets, 2-8
Pieces and Patterns, 5-9
Primarily Bears, K-6
Primarily Physics, K-3
Sense-Able Science, K-1
Spring Into Math and Science, K-1

* Spanish supplements are only available as downloads from the AIMS website. The supplements contain only the student pages in Spanish; you will need the English version of the book for the teacher's text.

Spanish Edition
Constructores II: Ingeniería Creativa Con Construcciones
   LEGO® 4-9
   The entire book is written in Spanish. English pages not included.

Other Science and Math Publications
Historical Connections in Mathematics, Vol. I, 5-9
Historical Connections in Mathematics, Vol. II, 5-9
Historical Connections in Mathematics, Vol. III, 5-9
Mathematicians are People, Too
Mathematicians are People, Too, Vol. II
What's Next, Volume 1, 4-12
What's Next, Volume 2, 4-12
What's Next, Volume 3, 4-12

For further information write to:
AIMS Education Foundation • P.O. Box 8120 • Fresno, California 93747-8120
www.aimsedu.org • 559.255.6396 (fax) • 888.733.2467 (toll free)

Duplication Rights

Standard Duplication Rights

Purchasers of AIMS activities (individually or in books and magazines) may make up to 200 copies of any portion of the purchased activities, provided these copies will be used for educational purposes and only at one school site.

Workshop or conference presenters may make one copy of a purchased activity for each participant, with a limit of five activities per workshop or conference session.

Standard duplication rights apply to activities received at workshops, free sample activities provided by AIMS, and activities received by conference participants.

All copies must bear the AIMS Education Foundation copyright information.

Unlimited Duplication Rights

To ensure compliance with copyright regulations, AIMS users may upgrade from standard to unlimited duplication rights. Such rights permit unlimited duplication of purchased activities (including revisions) for use at a given school site.

Activities received at workshops are eligible for upgrade from standard to unlimited duplication rights.

Free sample activities and activities received as a conference participant are not eligible for upgrade from standard to unlimited duplication rights.

Upgrade Fees

The fees for upgrading from standard to unlimited duplication rights are:
- $5 per activity per site,
- $25 per book per site, and
- $10 per magazine issue per site.

The cost of upgrading is shown in the following examples:
- activity: 5 activities x 5 sites x $5 = $125
- book: 10 books x 5 sites x $25 = $1250
- magazine issue: 1 issue x 5 sites x $10 = $50

Purchasing Unlimited Duplication Rights

To purchase unlimited duplication rights, please provide us the following:
1. The name of the individual responsible for coordinating the purchase of duplication rights.
2. The title of each book, activity, and magazine issue to be covered.
3. The number of school sites and name of each site for which rights are being purchased.
4. Payment (check, purchase order, credit card)

Requested duplication rights are automatically authorized with payment. The individual responsible for coordinating the purchase of duplication rights will be sent a certificate verifying the purchase.

Internet Use

Permission to make AIMS activities available on the Internet is determined on a case-by-case basis.

• P. O. Box 8120, Fresno, CA 93747-8120 •
• permissions@aimsedu.org • www.aimsedu.org •
• 559.255.6396 (fax) • 888.733.2467 (toll free) •